Beyond Command and Control

Leadership, Culture and Risk

Beyond Command and Control

Leadership, Culture and Risk

Richard Adams
Christine Owen
Cameron Scott
David Phillip Parsons

CRC Press
Taylor & Francis Group
Boca Raton London New York

CRC Press is an imprint of the
Taylor & Francis Group, an **informa** business

PHOTO CREDIT: Eucalyptus Leaf collected during BCC Simulation exercise, Mt Macedon, 2014 by Christine Owen

CRC Press
Taylor & Francis Group
6000 Broken Sound Parkway NW, Suite 300
Boca Raton, FL 33487-2742

© 2017 by Taylor & Francis Group, LLC
CRC Press is an imprint of Taylor & Francis Group, an Informa business

No claim to original U.S. Government works

Printed on acid-free paper
Version Date: 20161125

International Standard Book Number-13: 978-1-138-70883-9 (Paperback)

Visit the Taylor & Francis Web site at
http://www.taylorandfrancis.com

and the CRC Press Web site at
http://www.crcpress.com

Contents

Preface

Between 2010 and 2014 the authors collaboratively taught courses in an advanced professional development leadership program at the Australian Emergency Management Institute. This book emerged from that happy association. Beyond their constructive partnership, this text reflects the authors' shared conviction that leadership is most helpful when it enables others to do their best.

The text also reflects the interest of the many practitioners with whom the authors have been privileged to work. These people have enriched the text beyond measure. Their expertise has been in all manner of fields, including aviation, medicine, seafaring, marine salvage, offshore oil and gas, firefighting, critical infrastructure, the police, humanitarian and health services and in non-governmental organisations. Many were unified in their support for the book's main themes. Others struggled with the concepts. The conversations that emerged helped us to articulate these principles better.

The Institute that sponsored this program closed its doors in 2015 as a result of federal government budget cuts. Although there are many who remain committed to continued capability development in the sector, the type of program offered is yet to materialise in another context. This book has been written, in part, to give voice to the key intentions of that program and so that others may also benefit from the many rich conversations that led to the points elucidated in this book.

This book argues that the biggest challenge facing organisations today is not a shortage of bureaucracy. Rather, it is a dearth of leadership. The challenge facing organisations today is a shortage of individual responsibility. Organisations fall short when senior people come to be swellheaded, when they fail to lead by example and when they fail to engage with others in pursuit of collaborative problem solving. Informed by research and encouraged by people who do real jobs, this book is for those who seek to bear the weight of leadership and to do so with a true and ethical heart.

Authors

Richard Adams, the recipient of an Australian Fulbright Scholarship to Yale University, earned doctoral, master's, and first class degrees from the University of Western Australia, and a master's degree from the University of New South Wales. His professional interests and experience lie in ideas of leadership, institutional culture and risk. He is a researcher at University College, the University of New South Wales.

Christine Owen is a researcher with a focus on organisational behaviour and learning. Christine has established a growing reputation as a human factors researcher and facilitator within emergency management. Her research investigates communication, coordination and teamwork practices in high-technology, high-intensity and safety-critical work environments.

Cameron Scott is the National Network Emergency Response Manager at the National Broadband Network (NBN). Cameron has had emergency management roles in the state and federal governments, including the Department of Economic Development, Jobs, Transport and Resources in Victoria and the Australian Emergency Management Institute within Emergency Management Australia. He served as a sworn officer with the Western Australia Police for more than 13 years, working in the emergency management and counterterrorism sections, with responsibility for emergency planning, capability and emergency response. Cameron is also an academic at Charles Sturt University, where he coordinates units in the Bachelor of Emergency Management program.

David Phillip Parsons has thirty-eight years of experience in the emergency services industry. He is currently a Senior Emergency Management Adviser with the New South Wales Department of Industry and an Adjunct Professor at Queensland University of Technology Centre for Emergency and Disaster Management and Charles Sturt University's Graduate School of Policing and Security. In addition, he is a Fellow of the Business Continuity Institute and the Australian Institute of Emergency Services.

1

Introduction

1.1 The Primary Purpose of This Book

This book has one primary purpose: to advance the understanding of leadership beyond the inherited myths and traditions of command and control. It advances a collaborative model of leadership, arguing that the real power in leadership is not the power of one over others, but the power of the collective; the power leaders build with others in a joint effort.

Offering case examples, this book looks at the *pivotal moments*, episodes when leaders take judgements and act decisively. Beyond considering that one conclusive moment, we also learn about the cultural substance which leaders must establish. This book explains the things leaders should know and do to set the foundation stones of success. And we learn about the importance of communication between agencies and institutions, and between the pillars of government bureaucracies.

Illuminated by accessible and useful references to theory, the lessons in this text are practical and easy to appreciate. Many of the case examples that underline and illuminate arguments originate from people involved in emergency management in Australia, but the resonance and application of the lessons are wide ranging and they can be heard and understood by everyone.

1.1.1 Audience

This book will appeal to people who recognise that for those who are trusted by their fellow men and women, leadership is a privilege. When people are trusted to lead, they bear an obligation to lead well. This book puts forward the idea that leading well is informed by the concept of *power with*, not by the idea of *power over*. These phrases intuitively indicate this book will appeal to those who recognise that people in junior or supporting roles are not subservient.

Informed richly by research, this book will be valuable to people who lead in high-risk, high-stress, time- and safety-critical and high-consequence environments. It will be equally important to those who need to find creative solutions to complex and novel problems. When the crunch comes, conscientious people will want to lead responsibly and well. This book is for them.

1.1.2 Contesting the Myths of Leadership

In this book leadership is less about order-giving and more about human interaction. This book does not follow the well-worn path of established convention. It does not draw inspiration from management schools. It is not a book for bosses or bureaucrats. This book will not appeal to people who think the part of subordinates is to do as they are told.

This is a book for people who understand that the challenge facing organisations today is not a shortage of bureaucracy. It is for people who recognise that organisations depend more on the conscientious, responsible participation of high-quality employees than on big shots in swivel chairs.

This book questions myths concerning leadership, which are almost universal. The myth of leadership is the myth of elitism and seniority, the myth of personage and personality, the myth that some people are special, protected and entitled to behave in ways which are not conventionally acceptable.

These myths owe their provenance to the cultural misbelief that leadership is charismatic and exceptional. This error, which finds expression in the doctrines and assumptions of command and control, has a very wide currency and resonance. For example, the mandarin stalking the corridors of power exemplifies Western thinking about leadership – or at least the mainstream Western delusion about leadership.

The global cult of the extravagantly paid celebrity Chief Executive Officer (CEO) points to the seeming undisputed general acceptance that some people have a captivating magnetism and relevance the rest of us do not have. This hypnotising charisma is apparently enough to justify grossly unequal salaries and privilege. Similarly, in sport, vulgarian exhibitionism reflects the popular heresy that talent is an excuse.*

In the twentieth century, captivating magnetism in the person of a dictator convinced a nation to scandalise the conscience of humanity. We would like to think we have come a long way since the Holocaust – but in many ways, we have not. Highly cultured and educated Western societies often turn a blind eye to the misdeeds of politics and the transgression of big names. Commissions Against Corruption routinely find evidence of gross malfeasance by prominent people. And, just as routinely, rhetoric and celebrity seem to avert the criminal proceedings and prison sentences which would await lesser citizens. Similarly, the patriarchs of the Church – the princelings, the cardinals, the bishops and the other prominent men – appear too

---

* Regrettably, at the vernacular level, most poor behaviour in sports is improper behaviour by men. Equally regrettable and idiomatic, most extravagantly paid CEOs are men. The unlikable behaviours which cliché these people are described informally as 'masculine'.

big to fail. Eminence and seniority seem insurance against prosecution for the individual and collective failure of these so-called shepherds to protect young children entrusted to their care, though some steps have been taken to address this issue.

The Australian lawyer, journalist, author and political commentator David Marr wrote a devastating article on this subject in the 26 February, 2016 issue of *The Guardian*. Offering the case of Cardinal Pell testifying his ignorance in the Australian Child Abuse Royal commission, Marr asks us to wonder how people could be believed when they claim to have never read that particular letter or this report, when they claim to have been out of the loop, when they claim no one warned them, no person complained. Marr asks us to wonder why we believe senior people who claim to have risen through the ranks in a state of complete ignorance, unsmeared and unaware of systematic cover-ups. Marr asks us to challenge the prevailing myth of leadership.

This myth is the make-believe of notoriety, the bluff of seniority, the confection of eldership and precedence. The myth is cashed out in practice when senior elites are excused the obligation of policing themselves and their friends, when ordinary people are given kudos for assertive and self-confident behaviours.

The leadership myth presumes power and the right to command, to control, to dominate. There is the contained view that leadership is elevated and high. The characteristic upshot is the undeclared idea that some people are better than the rest of us. This myth is wrong. But it persists. Universally unsound, the leadership myth is alarming.

In safety-critical domains the leadership myth is outright dangerous. In the view of this text, the safety-critical domain is one in which people interact with each other, typically in the face of significant time pressures and often with high technology. Very often, the safety-critical domain is structured by complex regulation. Placing a special emphasis on the safety-critical domain, this book contests the leadership myth. This myth is the false hope in command and control, the fantastic notion that a loud voice, extroverted personal sureness and a talent for order-giving will be adequate.

In safety-critical domains, the mythologised legacy of the past can persist long after its uselessness has become evident. For example, aviation was dogged by machismo until National Aeronautics and Space Administration (NASA) exposed the positive correlation between abrasive confidence in pilots, accident and death. The surgeon with a God complex similarly dogs medicine. These colloquial illustrations carry the resonance of the rugged individuals who inspired the nineteenth-century *great man* theories.

The logical causation of the great man theory is complex and untenable today. Popular in the nineteenth century, the idea that great men were entitled to lead the rest of us drew its inspiration from great social and educational inequality. Cultured men could lead, and they did. As the West industrialised, the leading minds were those of educated elites: the brawn was on the shop floor.

1.1.3 Critical Ideas: Leadership and Command and Control

In this book, leadership is understood to be a matter of collaboration and relationship, and leaders are understood to be people who exercise constructive influence. Informed by the study of flight crews under stress, the ideas of leadership which are articulated in this book find application anywhere professionals interact with complex technical systems, in stressful and high-risk circumstances [35,36].

Emphasising the idea of constructive influence, this book separates ideas of leadership from ideas of organisational hierarchy or positional power. In a nutshell, the book draws a line between the official leaders and the real leaders. In this book, leadership is NOT positional authority exercised by seniors over juniors. In this book, seniority is about seniority – which is to say, seniority describes position, not leadership. We understand that some senior people do lead. But many senior people just direct, or they manage or they boss and bully. Many senior people are very far from leaders.

In this book, leadership is separated from ideas of rank. Leadership is understood to be an obligation borne by everyone, regardless of seniority in the organisational system, and regardless of social circumstance, religious faith or gender. Seniority has purely coercive institutional force: people comply because they have an economic need to work, and because senior people have the capacity to endanger the satisfaction of this need.

Leadership, by which is meant the constructive example of individual merit, is compelling, not coercive.

In this book, we call positional authority: *command*. The descriptive phrase 'command and control' is used to convey the sense of directive authority exercised by those in positions of seniority. Command and control describes the power of superiors *over* juniors, the power of managers or supervisors *over* subordinates. The success of command and control relates to task accomplishment, to meeting budgets and to checking boxes that bureaucracy wants checked.

To the person exercising command, it does not matter that the exercise of command is inspirational or not. To the person exercising command it does not matter that people are committed or not. Exercising command, nothing matters except that people do as they are told, and that tasks are accomplished efficiently; on time and on budget.

In a nutshell, the upshot of these definitions is to acknowledge ideas of seniority, and to separate ideas of seniority from ideas of leadership. Leadership describes ideas of individual merit and responsibility. Leadership is concerned with constructive personal influence. In this book, the practice of leadership is understood to be more collaborative than directive. The best leaders will be those who work with those around them, and who secure critical background conditions, building a culture or context within which others can do their best.

The effect of leadership is in the engagement of people. This is critically significant, especially when task accomplishment depends, not on managerial direction, but on the interaction of people with each other and with technical systems or complex regulations. This sort of interaction is typical in high-risk, safety-critical environments.

This book does not argue against hierarchy. But it does argue against the negative effect of hierarchy, which is to stifle the engagement of people, to suppress their energy and innovative spark. At its worst, hierarchy presumes all the answers are at the top, and that the old ways are best.

This book recognises that the people who work together in organisations occupy roles which have different places in the hierarchy. Some people are senior. Some people are junior. But this book puts forward that leaders will recognise a fundamental equality. This means leaders will recognise each individual as an independently responsible citizen. As citizens, all people are equal, each entitled to a single equivalent vote, and to equal justice under the law. And, as citizens, all people in an organisation are entitled to fair treatment.

This sense of equality is contained in the phrase *power with*. As a prevailing theme in this book, the idea of *power with* recognises that citizens work together in the organisational system.

In this book, the word *system* is used to describe a totality, where the whole is greater than the sum of the parts. As an analogy, the system is like the human body. In the human system, neither the head nor the heart is more important than the other. The head and the heart each depend on the other to function properly. So it is in the organisation, where each person must play his or her part responsibly. Each person is dependent on all the others. And in this system, people share power with colleagues, rather than exercising power over them.

For the organisational system to work well, leaders must recognise the equality of people, and they must enable the sort of critical human interaction which is often made difficult by institutional hierarchy. This book looks beyond organisational hierarchy. But it does not claim hierarchy is immaterial. Nor is this book insensitive to the very special responsibilities people bear as a consequence of their place and role in the hierarchy. But it does claim that the very conspicuous parts of the organisation's hierarchy can make us insufficiently attentive to elements of human interaction which transcend the formal structure.

In a nutshell, this book looks to ideas of responsibility, duty and obligation that attach to people rather than to rank. For leaders it is important to inspire and enable sensible, truthful, respectful communication, especially in organisations with traditional hierarchies. This sort of human interaction and exchange can be made difficult by an overprominent bureaucracy.

The unconstructive effects of hierarchy are amplified by the tendency of organisations to become bureaucratic. Bureaucracy is incompatible with leadership, which is a human quality, and contrary to fluent and constructive human interaction. In the bureaucratic culture, the human response tends to be numb and chicken hearted because, in bureaucracy, individual responsibility is valued less than rule following. In bureaucracy, people tend to be insufficiently attentive to *what's right* because they are inclined to be engrossed with trivial administrative rules and avoiding blame.

When organisations are excessively bureaucratic, there is a fetishised attention to minutia and detail. Official writing, for example, comes to be evaluated, not on the basis of clarity and relevance, but on the basis of margins and tabulations, and people get on because they are infatuated with the immaterial detail of the organisation's writing manual. Neatly set out, bureaucratic writing tends to be long-winded and ambiguous. The indecisive writing reflects the bureaucratic habits of risk avoidance and micromanagement.

Leaders should resist these pressures. Leaders should exert only so much authority as to ensure tasks are safely and professionally carried out. Positive control should never be excessive, and always respectful.

This remark points to the habit of bureaucracies to centralise control. Centralisation describes the bureaucracy's obsession for systemisation and the concentration of power. In practice, it means that information from the seat of events is passed upward to headquarters, which issue direction. This bad habit dissolves the autonomy of the individual on the ground and, as Jim Storr observes, is fundamentally unconstructive because 'The amount of information passed between a group of people increases roughly with the square of the number involved (a consequence of many-to-many information strategies), while the ability to deal with it increases only linearly' [81, p. 126].

In a nutshell, bureaucratic red tape routine operates to stifle leadership, which is understood to be the art of exercising constructive influence. In a bureaucracy, people tend to become good at issuing orders and reading spreadsheets and Gantt charts – but they also tend to be inept leaders. They lose the sense of human engagement, human initiative, creativity and responsibility. Bureaucracy is the antecedent of red-tape hierarchy and managerialism, but not leadership. This book offers a new way of thinking.

1.2 The Important Contribution of This Book

This book makes an important contribution because it separates ideas of leadership from ideas of formal command, control and organisational structure. It accepts that ideas of leadership have derived much of their impetus from cultural forms, and from the bureaucratic organisational conventions which emerge from the cultural forge. This book challenges these inherited forms.

Recognising hierarchy as critical and central to the efficient functioning of complex institutions, this book does not argue for doing away with command-and-control hierarchies. Rather, this book identifies the difference between formal or official leadership, and real leadership. In this book, a line separates ideas of positional authority and formal command and control from ideas of leadership.

The focus of this commentary is on leadership as an expression of collaborative human relationship. Recalling the eminent Western Australian General Sir John Hackett, leadership at its best approaches a *complete fusion* of the leader with the greater whole of the team [32].

1.3 Structure of the Book

The book proceeds as follows. Chapter 2 addresses the idea of unconstructive power. In Chapter 3, the alternative idea of *power with* is set out. Having made the case for a collaborative model of leadership the book then proceeds to discuss the key elements needed for collaborative leadership. These include creating conditions in which followers are responsible and play their part (Chapter 3); facilitating effective communication patterns (Chapter 4); establishing shared situation awareness (Chapter 5) and effective decision making (Chapter 6); and establishing conditions to support error management (Chapter 7). The book concludes with a discussion about the obligations to which leaders must commit (Chapter 8) and with a synthesis of the key ideas (Chapter 9).

1.4 Conclusion

This book proposes a model of leadership which is more collaborative than directive. It distinguishes positions of formal responsibility and power from leadership, which is understood to be the art of exercising constructive influence. Understood like this, leadership is compelling, not coercive.

2

Unconstructive Power

Humility must be the portion of any man who receives acclaim.

Eisenhower
Guildhall Address, 12 June 1945

2.1 Introduction

In a nutshell, this book argues against unconstructive power, which is a prominent feature of command-and-control models of leadership. It argues that miscommunication has a devastating effect in tightly coupled systems, where the decisions of one component have great implications for other elements within the system* [72,86]. This book advances a collaborative model of leadership. Doing so, it asks us to reconsider the stale sayings and the unquestioned assumptions which alibi the failure of old guard authority.

This chapter points to the failure of established and inherited ideas about leadership. It also sketches ideas which will be taken up and developed more fully later in the book.

2.2 The Failure of Inherited Ideas of Leadership

At a time when democratic societies around the world are confronted by the challenges of climate change, economic weaknesses and extremist violence, failure is not attributable to the resilience, the energy or the humanity of the people. The failure of democratic societies is the failure to find political and social vision. Decay follows from the failure to unite people in the coactive enterprise of nation building and social endowment. The failure is the failure of leadership. The failure of democratic societies is the persistence of

---

* The idea of tight coupling comes from complex systems theory. More information can be found in Sagan (1993) and Weick (1993).

worn out assumptions that the leader knows best, that the senior people have all the answers. But this is demonstrably not the case. For example, dismayed by the squandering of blood and treasure in more than twenty years of fruitless Middle East conflict, citizens of the great liberal democracies need look no further than their political and military leaders to find the culprits.

More wooden than the First World War donkey generals, modern politicians and military senior officers have expected us to trust them. They have told us what is best. They commit our children and the wealth of our nations. But they cannot tell us why. And they do not win. And when they fail, they can say nothing better than 'more money please'. Committing the nation to war with neither good strategy nor good reason, these people presume senior rank and public position is defence enough for precipitant decisions which, a long way from transparent, commit future generations to ostensibly endless unwinnable war.

The same unfounded presumption that senior people know best is evident in the way emergency services agencies disregard the intelligence available from social media. The significance of this discussion is in the deep, pervasive and the mistaken association of leadership with power and celebrity (Box 2.1).

Leadership in government and in other spheres of public life, which has no basis other than seniority, has lost legitimacy. Leadership depends on much more than precedence and position. Leadership demands character and competence. Contemporary institutions, the products of modern societies, involve and depend on people who are well educated, well informed, connected and questioning. Leaders whose thinking about leadership is

BOX 2.1   SOCIAL MEDIA AND DISASTERS

The use of social media during disasters to both disseminate information and obtain intelligence first came to the fore during the latter half of the first decade of the twenty-first century.

Many traditional emergency managers resisted the use of this technology as media that provided frivolous, useless information. Often an explanation of 'Why would I want to know what people had for breakfast?' was provided as a reason for not utilising these tools.

Also of concern was an apprehension about 'losing control' of information or public messaging.

The Queensland Police Service embraced the use of social media during the Queensland floods of 2011 and used it extremely effectively to disseminate information and effectively 'myth-bust' rumour and misinformation. Their effective use of social media placed them in a much stronger position as a trusted source of information in future disasters. [9]

dominated by shop-worn ideas of dominance, will underwhelm the people on whom they depend.

2.3 Leadership Is Not a Position

Leadership is often associated with prominent positions and dominant behaviour. But constructive behaviour, which is critical to successful leadership, is often not properly understood.

A powerful illustration of this claim was offered at the conference *Resource Management on the Flightdeck* sponsored by NASA in 1979 [3, pp. 4–6]. Research presented at this conference 'identified the human error aspects of the majority of air crashes as *failures of interpersonal communications, decision making, and leadership*' [3,45,62,71]. This research sparked significant further enquiry, and sundry broadly applicable ideas.

In a nutshell, this workshop exposed and confronted the inability of junior pilots and senior pilots to work cooperatively. Accepting the many explanations for this, one reason stands out – it is the excessive assertiveness of senior pilots, which makes it impossible for junior pilots to play their proper part (Box 2.2).

The issue of dominant seniors and intimidated juniors is ubiquitous and difficult to address, because the personal habits and behaviours that imply dominance are just that – personal, contextual and difficult, if not impossible, to define as specific behaviours, and woven tightly throughout the organisational fabric.

Robert Helmreich's postdoctoral research into aviation safety at the University of Texas underlines the difficulty and the complexity of what has come to be known as human factors [35]. Helmreich writes:

> The culture of pilots is a strong one – exemplified by the rugged individualism vividly portrayed by Tom Wolfe in *The Right Stuff*. In addition to having great professional pride, many pilots strongly deny susceptibility to stress – they are unwilling to acknowledge that fatigue and sudden danger can dull thinking and slow response times. This sense of invulnerability can manifest itself in a desire to play the role of the white-scarfed lone aviator battling the elements. (p. 66)

Helmreich may depict an extreme – perhaps a myth. However, the allegory carries significant and pervasive *cultural* influence. The invulnerability and bravado, which Helmreich attributes to pilots, is not exclusive to them. Helmreich captures much of the pretension which is associated with leadership more generally, and which has an equally conspicuous expression in many other professions.

BOX 2.2   MANAGING OVERBEARING SENIORS

In one US hospital many patients died as a result of the power differential between surgeons and nurses. Nurses maintained a white board showing each surgeon's name – blue meant nice, red meant jerk and black meant whatever you do, don't contradict them or they'll take your heads off. Nurses watched as patients died or had incorrect surgery. The hospital was fined hundreds of thousands of dollars and eventually installed video recording in the operating theatres to monitor surgeon behaviour. [20, pp. 154–181]

The point to sheet home is the mistake of associating leadership with power, position and perfection. These words speak to stereotypes and myths, which represent themselves throughout organisational fraternities and subcultures. They are myths which catch senior people. Immersed in a culture of influence and prestige, senior people can come to believe their own press and arrive at the mistaken conclusion that they are immune to stress and mistakes [37].

More visibly, myths of power and supremacy make it obviously difficult for juniors to collaborate with seniors. Describing the danger of excessive power gradients, Edmondson [22] writes:

> Power differences…intensify the interpersonal risk faced by (people) who wish to speak up with ideas, questions, or concerns. Leader actions thus may affect whether or not people are willing to speak. The interpersonally safe route is to remain silent, but this poses technical risk…. Not speaking up can protect individuals but harm the team (or the organisation as a whole and magnify operational risk). [22, p. 1420]

The hazard of an unconstructive power gradient is amplified by the innate structure of organisations, and by the natural inertia of organisations. This means most organisations have an hierarchy and, in most organisations, people tend to be unconcerned by people who take advantage of hierarchical structures in a clumsy and overbearing way. People tend to brush away concern with remarks like: 'it's always been that way', or 'it does not matter what organisation you talk about, there is always someone at the top' (Box 2.3).

But remarks like this ignore a potent truth: the unconstructive, clumsy, or overwhelming use of formal authority is not merely an unpleasant behaviour, but a risky one. This truth was made very plain in 1979 when the National Transportation Safety Board published a report on an investigation into the 1978 loss of a United Airlines aircraft [1]. This report identified the improper dominance of the aircraft captain and the equally unacceptable timidity of junior crewmembers as causal factors in the loss [62, p. 120]. The report observed that 'the stature of a Captain, and his management style, may exert

> **BOX 2.3   THE IMPORTANCE OF COLLABORATIVE CULTURE**
>
> National Transportation Safety Board data show that 73% of airline incidents occur on the first day a crew fly together, and that 44% of these incidents occurred on the very first flight. The data showed the importance of working together to develop familiarity and effective communication and knowledge of each team member's strengths and weaknesses. [33]

subtle pressure on the crew to conform to his way of thinking. It may hinder his interaction and force another crew member to yield his right to express an opinion' [1, p. 27]. The report noted the danger of this sort of dominance, and commented that 'the first officer's responsibility is to monitor the captain (and to) provide feedback for the captain. If the captain infers from the first officer's actions or inaction that his judgment is correct, the captain could receive reinforcement for an error or poor judgment' [1, p. 27].

The nub of this report is this: some senior people will use their senior positions to intimidate their juniors and to get their own way. But more than an unpleasant behaviour, it is unsafe, because intimidated juniors are put in a position in which it becomes psychologically impossible for them to speak and to correct the errors or deviations of seniors.

When intimidated juniors find it psychologically unsafe to correct the errors of their seniors, accidents may not occur. To be plain: accidents may not be the direct result of a junior remaining silent. But, even in the absence of accidents, risk is elevated when juniors are uncommunicative.

2.4 Leadership and Risk

Risk is diminished when seniors make it possible for juniors to correct mistakes and deviations from best practice (both their own mistakes and deviations, and those of the senior). But this sort of human responsiveness can run counter to the acculturated cultural narrative about leadership.

Describing the loss of the United Airlines aircraft, the National Transportation Safety Board (NTSB) described the improper dominance of the aircraft captain. But the report also described an inventory of foreseeable behaviour. This means that the dominant aircraft captain was an acculturated expectation. Dominant behaviour by the aircraft captain was part of not just the cultural structure of the airline, but also of a wider cultural fabric; it was part of the pervasive cultural myth. The aircraft captain conformed to pervasive social norms, to a cultural model that was so strong and influential that, for any aircraft captain to *not* exude an almost overconfident 'command presence' would have been very difficult.

Loath to appear 'weak', or 'incapable', aircraft captains in 1978 – when the NTSB identified dominant behaviour as causal in the loss of an aircraft – would have conformed to social expectations which would have made it difficult for them to take the advice of junior crew.

This is because, like people everywhere, pilots respond to the normative weight of social cues and assumption. An assumption is something that is taken for granted rather than explicitly verified [7]. Assumptions are also an important and essential part of human activity. They are usually considered in relation to what individuals think and do. But assumptions may also be shared among social or professional groups, and in this regard, assumptions can take on a fictional or myth-like character [7].

This book argues against assumed ideas of power, command and control. It argues for a collaborative model of leadership, and asks us to consider how institutional structures which emphasise the strict seniority systems implied by organisational hierarchy can overshadow and overpower the importance of responsible individual judgements and partnership.

This book does not argue that organisational hierarchy and formal seniority structures *will* result in poor performance. But formal hierarchy *might* contribute to poor performance. In the perspective of this book, a formal seniority structure would contribute to poor performance and to elevated risk when it amplifies the power gradients and makes it difficult for juniors and seniors to engage collaboratively. When seniors are helped by institutional structures to appear senior, distant and intimidating, then formal institutional arrangements are likely to have become unconstructive.

Using the term unconstructive, this book means that the problems which follow from poor human dynamics will generate a *latent* danger. In a term taken from Professor James Reason, the idea of a latent danger describes an acculturated exposure to error. But the risk, being acculturated, is unseen – it is tolerated or accepted as part of the organisational fabric. Such risks are realised for what they are only in the investigation of an accident. The latent error contrasts with the error of judgement which, made in the moment and proximate to consequent events, is called an 'active error' [69, p. 173].

This book argues against old-fashioned ideas of dominance. It argues for an expansive and more collaborative idea of leadership, a model which accepts that every person has a role to play, and that it is the part of the responsible leader to help juniors and subordinates play their part as well as they might. This claim underlines the importance of shared situation awareness, and the value of an environment in which everyone feels able to question assumptions and actions and indeed accepts the responsibility to do so, and in which supportive professional responses are the norm: in other words, *what is correct*, not *who is correct*.

Making these points, this present book makes an important contribution in the emergency management domain, as the professional practice of emergency management evolved from the military and retains several conventions and assumptions inherited from the armed forces.

Few of these hereditary ideas are more conspicuous or significant than the idea that leadership is a practice attached to command and control.

2.5 Command and Control Is Important

Without clear direction, and without explicit accountabilities, organisational performance will be a long way from efficient. But, when they are overreliant on commands and insufficiently attentive to institutional culture, seniors can fall afoul of the unconstructive false impression that power works and that the success of leadership is measured exclusively in the terms of task accomplishment. But the façade of efficiency is just that – a façade, a false hope and a false impression. People can focus on nothing but efficiency, and fail to lead. In a significant way, leadership is about more than direction giving. Leadership is about engagement and involvement.

On the part of juniors, a *command climate* can be intimidating, and it can give a false impression. This false impression is the idea that conformity with hierarchical norms is sufficient to fulfil duties and obligations.

2.6 Collaborative Culture

Looking beyond command and control, this book asks leaders to be mindful of their larger responsibility to build a communicative, collaborative culture. Focused intently on getting a job done, people can rely on positional power and neglect more critical responsibilities. In the perspective of this book, these responsibilities include the responsibility to collaborate, to foster a *power with* culture. This is an obligation for two key reasons. One, collaboration ensures all cognitive resources are brought to bear solving problems. Two, collaboration offers individuals the opportunity to be engaged in and fulfilled through their work. Collaboration means that juniors are not treated as subservient – which regrettably, in many cases, they are.

Stakeholders – a regrettably clichéd word – are shareholders, and often participants in the enterprise. They have a vested and significant interest in organisational means and outcome, and should be engaged as associates and contributors. Their engagement is a responsibility of leadership. How does a leader engage people? Apart from by personal charisma, the engagement

BOX 2.4   GARNERING HELP IN A CRISIS

In response to the February 2011 Christchurch earthquake, Sydney Water's Managing Director Dr Kerry Schott encouraged the company's emergency manager David Phillip Parsons to help Christchurch complete emergency repairs on the sewer system. Dozens of telephone calls to senior executives at Water Infrastructure Group, Veolia, GMA Environmental Services, Interflow, Kembla Water Tech and Barry Brothers saw a task force of trucks and personnel join Sydney Water in going to Christchurch. David used preestablished relationships in Civil Defence and Emergency Management New Zealand, Christchurch's provider City Care and Australian emergency services to facilitate the largest ever international deployment of water company staff.

and involvement of others is by securing the background conditions within which people might thrive and do their best – in other words, to build what might be called a constructive organisational culture.

The practical advantage of these claims is in moments when people work together in high-stress, high-risk and time critical circumstances (Box 2.4).

Claiming the importance of collaboration and offering examples of combined effort and partnership, this book takes advantage of lessons drawn from the performance of flight crews under stress and from other high-risk environments such as the bridges of ships and the control rooms of nuclear and petrochemical plants. In these places, the interaction of professional people with each other and with a complex control system is enabled by the deliberate practice of collaborative skills [35,37]. People think deliberately about the conventions and the patterns of their interaction. They build the procedures and the professional language, which will help them to work effectively. And they do this among themselves, in partnership and in collaboration with leaders. Effective habits and conventions cannot be ordered. People may comply with orders, but they will not commit to them – and collaboration requires commitment.

In some ways the consequences of human interaction are more important than formal organisational architecture. Edmondson writes:

> A team in a fast-paced action context…might have a clear goal (putting out fires, saving patients' lives, landing an aircraft), the right mix of experience and skills, adequate resources, a task that calls for teamwork (and) structures that support effectiveness – yet still suffer a devastating breakdown in coordination due to miscommunication, interpersonal conflict or poor judgement in the heat of the moment. Weick (1993), for example, attributed a tragic fire-fighting failure to a collapse in mutual sense-making in a group that had in place (the) basic structural supports for effectiveness. [22, p. 1420]

Edmondson speaks of collaboration, which is critical to a safety system. She recognises properly functional, properly insightful and engaged people make safety work. But Edmondson speaks of the single moment and the single place. She references the fire ground, the hospital theatre and the aircraft cockpit.

But leadership is more complex than in the determinate, single moments and explicit evolutions to which Edmondson refers. Leadership, beyond the hot moment and the single agency, has a critical effect in the long-term preparation for response and in the immediate lead-up to response. As well, leadership shapes the relationship of people and agencies to each other. In this way, leadership helps to alleviate the misjudgement and disorientation, which, in composite teams or multiagency groups, can be camouflaged by behaviours known to arise from the phenomenon of swift trust.

This is a perspective on leadership that recognises risk and safety-critical systems often extend beyond single institutions and agencies. Yet, the memes of overpowering authority and bureaucratic micromanagement which hallmark leadership delusions foster an unconstructive culture of competitive empire building. People who fall for the false need to seem more important and more powerful than other figures drag their organisation down the same wrong path. The practical effect is an antagonistic rivalry among people, services and organisations, when cooperation and collective effort is so much more fruitful [23,52,54].

The upshot is that leaders need to make it possible and constructive for people from different organisations to work together with ambiguous and sometimes conflicting information, under conditions of duress, with equipment and resources that might be inadequate and in environments that might be dangerous.

This need requires effective communication and coordination. Leadership will not be successful when the leader is a monodimensional director and order-giver. Successful leaders will build a shared (and distributed) situation awareness.

2.7 Conclusion

This chapter identified the flaws of the prevailing mindset: that the leader is the person at the top and the person who knows best. In a related discussion, this chapter argued against the misrepresentation of leadership as direction-giving. The alternative course is in a collaborative model of leadership, a model in which leaders foster human interaction.

3

Power and Partnership

3.1 Introduction

Chapter 2 considered the idea of unconstructive power. This chapter adds to the conversation. The discussion in this chapter recalls the ideas of the pioneer theorist Mary Follett, who coined the transformative idea of *power with*, and drew a distinction between this and the traditional *power over* thinking. The chapter explains leading and following as a partnership or collaboration established on the cornerstone of responsibility.

3.2 Leadership Is a Cultural Idea

Asking us to reimagine, or to reinvent, our ideas of leadership, this book challenges significant deeply rooted cultural assumptions that pervade different parts of our lives. There is the influence of national culture, the influence of organisational culture and the extra distinct influence of professional cultures – for example, the culture of doctors, or pilots, police officers or firefighters.

What this means is: when professionals arrive at work, they carry with them the forming and inescapable influence of their society, of the organisation for which they work and the profession to which they belong. All of these influences can affect performance [25,35] (Box 3.1).

Foremost amongst the cultural ideas which affect performance is the idea of power. Interpreting power as a cultural idea, the work of Hofstede has been significant. Hofstede clarified the idea of power and the concept of *power-distance* in organisational, social and cultural systems [38–40]. Following Hofstede, power is defined as the capability of one organisation member to direct the behaviour of others. Power *distance* speaks to the balance of power, or the power *gradient*. This means, when one person has very little power, and another very great power, then there is a large power distance

BOX 3.1   KINGS CROSS STATION FIRE

The Kings Cross Station fire occurred in the London underground in
November 1987. The bureaucratic structure of the underground meant
that personnel operated in strict silos with distinct responsibilities. As
a result, initial clues to the growing fire were not exchanged between
staff on shift as they were from separate business units. 'Even at the
highest level, one director was unlikely to trespass on the territory of
another', an investigator would later write. That night thirty-one people
lost their lives. [20, pp. 166–175]

or gradient. This is significant because conspicuous power gradients inhibit
communication and the helpful flow of information.

Typically, in groups, members with less power defer to those with more
power, protecting themselves through self-censorship to avoid being
rejected or marginalised. This means: where there is a formal power difference, and when speaking up matters for performance, those with power
have an obligation to lessen the inhibiting effects of the power gradient [22,
pp. 1423–1424].

This book puts forward the idea that leaders are in a partnership with followers. In this book, leading is understood to be collaboration. Good leaders
share power with their followers. Following is understood not as passive and
subordinate but as a responsible engagement with leadership.

3.3 Power With

Coined by Mary Follett in her important book, *Creative Experience*, the idea
of *power with* foreshadowed Joseph Nye's concept of soft power. The depic-
tive phrasing *power with* speaks against the ideas of dominance and coercion
which are contained in the established discourse. Overshadowed by assumptions of *power over*, the established or traditional discourse is informed by
ideas of precedence and yielding. There is an implied polarity, a sense one is
powerful and the other powerless.

Ideas of *power with* are not ideas of equality. But *power with* is about connection and combination. Unlike *power over*, the basic ideas of *power with*
gesture to collaboration from which all parties benefit. Where *power over* is
a notion defined by inevitability that juniors will obey or face the consequences, the idea of *power with* is different. *Power with* is informed by the
sense of possibility and responsiveness. The *power with* idea contains the
human sense of unity and engagement (Box 3.2).

BOX 3.2   THE COLUMBIA DISASTER

The inquiry into the Columbia Disaster which killed seven astronauts heard this testimony from the Mission Management Team (MMT) chair:

Investigator: As a manager how do you seek out dissenting views?
MMT Chair: Well when I hear about them.
Investigator: By their very nature you may not hear about them.
MMT Chair: Well, when somebody comes forward and tells me about them.
Investigator: But, what techniques do you use to get them?

Apparently the MMT chair offered no response to this final question. [49]

The practical effect is this: in situations of unforeseen or atypical problem solving, the *power with* approach enables the organic, innovative responsiveness of the group. In situations where there is a standard operating procedure, and where specialised roles are allocated among a group, the *power with* perspective enables effective risk mitigation and lessons to be learned without fear or favour.

How so? Imagine a senior surgeon, or a senior pilot, or a senior officer about to make some mistake but prevented by the junior who speaks up. And imagine the wash up debrief, where seniority is not a factor, so that lessons might be learned without prejudice. These examples gesture to *power with*, a leadership system where rank has no part, but performance and responsibility do. The idea of a system is important.

The term 'system' is often overused or used incorrectly. The significance of the idea is illustrated by reference to the human body: what would you rather have, the head or the heart? Of course, the question is nonsense, since we cannot have one or the other. We must have each together; this is because the body is a system – it works only when it is interconnected. Each part has a specific function, and each part depends on the other elements. So it is in the case of *power with* leadership. The group is a system: leading and following is a partnership. The point is: while positional seniority or specialisation is important, so is the interaction and the connection of each part.

From the *power with* perspective, it is the cooperation of each element in a group which generates a sort of self-sustaining human energy or power. From the traditional *power over* perspective, the power in a group is a monodimensional directive power, which must either grow larger itself or diminish the power of junior people in order to sustain the relative power proportion or balance. This means, in a *power over* model, junior people can never properly flourish, because the risk of a junior overshadowing the 'master' is intolerable.

To enable a *power with* relationship we must consider how we can reduce our dependence relationship on *power over* habits. Mary Follett writes, 'Genuine power can only be grown; it will slip from every arbitrary hand that grasps it; for genuine power is not coercive control, but coactive control. Coercive power is the curse of the universe' [28, p. xiii].

These coactive *power with* relationships require *integration* – resonant with the general culture of democratic society [28, p. 209] – and *transparency*, consistent with routine expectations of accountability (Box 3.3).

In the view of this book, the directive, emphatic and assertive habits of command and control can have unfortunate effects *when they outrun ideas of integration*. Taking much from Mary Follett, this book argues that ideas of control should be tempered by ideas of integration – or *working together*. Follett writes, 'Enough has been said of domination whether obtained by show of power or use of power; unless we can learn some other process than that we shall always be controlled by those who can summon to themselves the greatest force' [28, p. 156].

The term 'integration' speaks to ideas of trustworthiness and relationship. The idea acknowledges the importance of constructive exchange between people. The nub is partnership, not compromise [28]. The difference is that, where compromise requires individuals to surrender important ground, partnership emphasises the construction of new and mutual solutions. Partnership which places the weight on common purpose, efficiency and innovation rests on the groundwork of transparency. The idea of transparency points to leadership which is concerned not with grabbing power, and not with exercising power, but with *evolving* power and building *power with* other people.

Follett describes how, in so many organisations, information is either irrelevant or incomprehensible or it is propaganda [28, p. 212]. People will come together only when there is a commitment to honest, accurate and useful information. The idea is to make the working unit a functional whole, united

BOX 3.3   POWER WITH INTERGENERATIONS

In 2008 Sydney Water introduced a methodology of having 3G teams in its Emergency Co-ordination Centre. 3G teams comprised staff representing three generations of staff – those in their 20s, 40s and 60s. This ensured diversity in approaches to utilising information and understanding societal attitudes and needs.

(Source: David Phillip Parsons – Manager, Emergency Management and Operational Continuity, Sydney Water)

as much by a purpose or goal as by shared human understanding and relationship [28].

In practice, collaborative leaders share genuine information and straightforward opinions. Doing so they build trust, understanding and collective will. The payoff from their transparency is in cooperative engagement, strong morale and reduced risk. This is because from shared awareness comes collective wisdom, practical innovation and a collaborative real-world response to risk. Advancing collaborative ideas of *power with* leadership, this book argues against excessively directive command-and-control approaches.

But this book does not argue against power. This book asks for a new orientation towards power, for a reimagining of power, for a subtler and more nuanced reading of power. The argument is for a constructive model of leadership, which derives its foundational strength from self-control. In a nutshell, this means, 'the more power I have over myself, the more capable I am of joining fruitfully with you and with you developing power in the new unit thus formed (between) our two selves' [28, pp. 189–190]. The nub is: leaders cannot command commitment and cooperation. At best, a leader can *command* compliance.

Constructive leadership recognises *the inclination of others to follow or not.* Thus, the influence of a constructive leader is nothing more than the example of a good person and the drawing power of such a person. A constructive, collaborative leader will facilitate the collective will of others. This means that a constructive leader will help to find the expression of a collective will. A constructive leader will help to put into words what people think and believe. Doing this, drawing together the countless different perspectives which arise among people in organisations, a leader will build a common narrative.

This common narrative will be a cornerstone of the institution's response to the world. The common narrative will make the *organisational will* accessible and examinable. This narrative, by concentrating and connecting the different thinking in a group, helps to make a group much more than a mere loose association.

A collaborative leader builds understanding, conviction and commitment: the collective wisdom which gives confidence, a sense of meaning, a sense of belonging, a sense of individual respect to people besides a sense of partnership and unifying morale.

3.4 Soft Power

This book takes much from Mary Follett, who, as mentioned previously, coined the idea of *power with*. Follett foreshadowed Joseph Nye, who advanced the idea of soft power [60]. Describing soft power, Nye describes a binary model – soft power OR hard power. Hard power is the coercive force of command. Soft power is the winning, persuasive influence of attraction. In effect, writes Nye, the practice of effective leadership requires hard and soft power to be appropriate in the context. Sometimes, there is no alternative except to be directive. On other occasions, direction might be quite the wrong thing to do.

Nye cites Dwight Eisenhower, who put the case for soft power, and who underlines the *power with* theme. Eisenhower said, 'leadership is an ability "to get people to work together, not only because you tell them to do so and enforce your orders, but because they instinctively want to do it for you... You don't lead by hitting people over the head; that's assault not leadership"' [60, pp. 309–310].

Eisenhower's remarks coincide with the idea of this book that coercion is very easily overplayed, and that when it is overplayed it can have a destructive effect. Certainly, coercive hard power can build barriers and resentments and completely undermine the ability of leaders to nourish partnerships and combination.

In a nutshell, leaders will need, on occasion, to use hard or coercive or directive power. But they should be very careful, and make sure the use of command and control is utterly suited to the situation.

3.5 Loyalty and Obedience

The tension between hard *power over* and softer *power with* approaches has a flow on to the ideas of loyalty and obedience. These ideas are important because leaders want those who follow to be loyal and obedient, committed and wholehearted. But this willing and enthusiastic spirit can be crushed by overuse of forceful *power over* command styles, which are an acculturated part of Western thinking about leadership.

Leaders who foster loyalty and obedience appreciate that loyalty is reciprocal, and obedience is not blind. Just as followers support leaders, so leaders support followers. Leaders foster a culture within which people can follow with moral confidence. This idea speaks to the claim that loyalty is cultivated by trust and mutual respect. This means that loyal people give honest opinions courteously and with good will. Loyal people take care not to damage the reputation or character of others. Loyal people appreciate that speaking poorly of others is not plain speaking or a courageous display of integrity. Speaking poorly of others is ill disciplined gossip which undermines cohesion and *esprit de corps*.

The idea of loyalty is closely connected to the idea of obedience. Obedience is not blind: obedience is uncompromising – but it is also nuanced and complex. People bear an obligation to obey. But no one ought ever be expected to be mindlessly, unquestioningly, unthinkingly or carelessly obedient. People ought always be expected to use judgement and initiative, to be conscientious and to be responsible.

In a nutshell, obedience does not mean downright mindless compliance. No leader should expect this sort of behaviour. Obedience must be characterised by responsible consent and good conscience. This means that people ought to obey direction, *unless* they are convinced that direction is wrong or unsound. And if this is the case, people have an obligation to speak up, and leaders have an obligation to listen. Leaders will be confronted by non-compliance. When they are, they must act fairly, with a mind to justice and to the difference between honest error and deliberate violation.

3.6 Responsibility

Leaders collaborate on the cornerstone of responsibility. Leaders understand that, alongside followers, they are partners in complex events in which obligations are mutual and overlapping. The idea of mutuality recognises responsibility is infinitely wide. This means that responsibility borne by one person does not dissolve or terminate the responsibility of other people. Leaders and followers bear the same obligation: to be responsible in the strict seniority system of the institution.

Two things follow from the claim that leaders and followers bear an equal expectation of personal responsibility: First, leadership is not domination. Second, following is not submissive. Sharing the responsibilities of partnership, followers have an obligation to be dependable, not docile. Responsible followers tell the truth and give honest opinions. Leaders do not expect pandering and work to foster the independent voice of others. Leaders expect, and they enable, followers to stand up for what's right. In short, leadership is about *collaboration with*, not *power over*. The days when diktat passed for leadership are over.

3.7 Conclusion

This chapter discussed the partnership between leading and following. The chapter explained the idea of *power with*, which was distinguished from *power over*. The need for forceful command styles was acknowledged, but the need for wisdom in the application of directive approaches was addressed. The chapter concluded with an overview of the concepts of loyalty and obedience.

4

Leadership and Communicating

As I would not be a slave, so I would not be a master. This expresses my idea of democracy.

Abraham Lincoln
From an address to an Indiana Regiment, 1 August 1858

4.1 Introduction

The preceding chapters have spoken against the unconstructive misuse of power. The discussion has offered a more collaborative leadership model. The idea of collaboration was illuminated by reference to the idea coined by Mary Follett of *power with* rather than *power over* [28]. Collaboration was also seen to derive impetus from the idea of soft power, which was made famous by Joseph Nye [60].

This chapter examines the ideas of leading and communicating. It presumes that communication is essential to collaboration, and thus to leadership.

4.2 Be Plain

In hierarchical organisations it can be terrifically difficult for juniors to raise questions and concerns with their seniors. And it can be very difficult for seniors to listen – or to *hear* – to properly acknowledge what is being said. Hierarchy, in other words, can get in the way of constructive human exchange. It takes the spirit and the sense of leadership to get over this bridge. This is a sense which recognises the question is not an insubordinate challenge, or an admission of incompetence: the question is the antidote to uncertainty and mistakes.

Having said that, there needs to be clear recognition of the difference between an appropriate question and defiant or fractious, uncivil backchat (Box 4.1).

BOX 4.1   MINDGUARDS AND THE BAY OF PIGS CRISIS

When Arthur Schlesinger Jr (Harvard historian and presidential advisor) raised concerns with Robert Kennedy about his brother's plans to invade Cuba, Robert said, 'You may be right or you may be wrong, but the President has made his mind up. Don't push it any further. Now is the time for everyone to help him all they can'. [43]

The crux: to exercise leadership, people must appreciate the importance of effective communication. People who wish to lead must be able to look beyond hierarchy. And they must be able to look beyond slick and meaningless clichés. Leaders will foster and enable meaningful exchange. Leaders will not allow the festering sore of cordial hypocrisy to undermine their organisation, nor will they allow disorderly, ungovernable discourtesy. Leaders will insist that people use their words, that they speak clearly, and openly and politely, and that they write plainly.

The importance of effective communication is demonstrated by the countless cockpit voice recordings which reveal the inability of first officers and flight engineers to alert aircraft captains to impending disaster, and the equally distressing inability of aircraft captains to hear questions and concerns.

Too often questions or concerns are not heard and understood as *professional* questions and *professional* concerns. Perhaps questions are seen as *personally* insubordinate or *professionally* irrelevant. That said, questions and concerns might very well be expressed in a *passive* or *digressive* way – lacking an essential and professional authority. A digressive or inert tone of this sort will make it awkward for people to detect the significance of a question or remark.

Voice recordings are significant: they reveal the importance of plain speech in critical moments, and they reveal how important it is for people to voice concerns and to resolve ambiguity. Less conspicuously, cockpit voice recordings *show how important it is to foster a culture of plain-spokenness and clear expression*. The logic is this: people face critical moments with the habits they have, not necessarily with the habits they need.

The time to foster critical habits is before the time of trial. This means: critical habits must be nurtured as part of the underpinning institutional culture. However, even though the successful resolution of critical situations depends very much on people voicing concern and resolving ambiguity, this sort of thing happens rarely, for many reasons. Sometimes, not raising or responding to an external cue can be because the cue was not recognised. However, more often, a lack of response is because social influences inhibit an effective response.

For example, in one study [50] of fire fighters on a fire line the reasons for not voicing a concern included

a. Fear that no one will listen
b. Pressure to remain silent for career concerns
c. Becoming distracted or complacent

Another study [6] identified leader behaviour as a significant factor, inhibiting the capacity of followers to speak up. Inhibiting leader behaviours included

a. Failing to test assumptions
b. Failing to look for countervailing views

This sort of leader behaviour does not seem causally disconnected from the reluctance of followers to speak up *in deference to perceived leader expertise.* This point is significant because, whilst confidence and mastery is constructive leader behaviour, there is probably an indefinably fine line between confidence and the sort of excessive sureness which gives other people the false impression that they are wrong and better off remaining quiet.

When a leader speaks with *measured* self-assurance, confidence, with conviction and with an unobtrusive sureness the leader will make it possible for others to share their own thinking and concerns.

A properly confident – but not an overconfident – leader will build the sort of integrated *power with* environment within which everyone can adjust to the circumstances and conditions. For leaders, a deliberate yet understated confidence will not overpower others. Leaders query; they question themselves and they question others. This means leaders never take it for granted that they are in the right, and they are ready to hear the opposite views of others. Quietly confident and inquiring, leaders build an involving and integrated environment wherein all people feel they have the capacity to behave responsively and intuitively to real-world events.

People who feel engaged, not coerced or overwhelmed, respond innovatively and gainfully. When leaders are not feared, people reorganise resources and actions as necessary. When leaders are understood – not feared – people will not feel constrained or compelled to abide by directions or instructions which no longer suit the state of affairs.

How do people feel engaged, not coerced? How do people come to innovate and use their initiative? When are leaders understood but not feared? The answer is: when leaders devote time to the advancement or development of others, and less time to criticism. When leaders invest in the success of others, then work comes to have meaning and personal value beyond a salary.

The point is: efficiency, improvement and safety are achieved through human processes and relationships. These things might be specified in regulations and in standing operating procedures, but their achievement depends on human relationships.

Why so? Because when leaders foster a sense of human connection – the meaning of this expression is also found in words like trust and respect – then people will share what they know; people will find the confidence to raise concerns about even the weak signals which preface an accident or failure.

In a nutshell: when leaders secure background conditions, which make it possible for people to speak up and raise concerns, then leaders make it more likely that a team will adjust, adapt and avoid error. When signals are left unaddressed – and especially when the organisational culture makes it impossible to address them – then the latent danger is the risk of larger cumulative problems and failures in safety.

4.3 Challenges to Effective Communication

Effective communication can be hampered by both inadvertent and intentional, human behaviours. Unintentional and intentional behaviours will be equally erosive of organisational function. But intentional poor behaviour tends to be bad mannered, and for this reason deliberate poor behaviour is doubly intolerable.

To correct unhelpful behaviour, and to guard carefully against unconstructive habits in themselves, leaders need to be aware of the unconstructive habits which undermine communication when people are listening and looking, when they are speaking and in general. When listening or looking, effective communication can be undermined by

   a. *Listening or looking with preconceived ideas*: The receiver will likely see or hear what is expected or desired – but not hear what is said or see what is presented.

   b. *Listening or looking whilst distracted*: Understanding requires conscious effort or attention.

   c. *Listening or looking when impatient*: This will encourage the receiver to put words into someone else's mouth, or finish his or her sentences, or to see what is not really there.

   d. *Overlooking body language*: Communication is more than just words; it is also signs, tone of voice, facial expression, gesture and so forth.

   e. *Listening with undue deference*: Being intimidated or unduly impressed by someone can lead to embarrassment or fear of asking, 'what are you saying?' or 'what do you mean?'

   f. *Being disrespectful*: Disrespect is an impediment to good communication.

When speaking, effective communication can be undermined by

a. *Not establishing the basis for understanding or a frame of reference*: Ideas make sense in context. Without a shared context or frame of reference there is no basis for understanding – people will not be 'on the same page'.

b. *Omission of information*: In not wanting to attract unwanted attention, or wanting to avoid questions, pertinent details may be left out – a partial truth may be told, and therefore the full truth may never be expressed or understood. Or, people may be left wondering about the intent of instructions for want of a significant detail.

c. *Inserting bias or opinion*: Personal prejudice or opinion may colour the effective communication of fact.

d. *Assuming messages depend only on words*: Tone, inflection, gesture and body language convey meaning alongside words. Meaning may be lost when these elements are ill chosen. In fact, meaning may be completely altered, when non-verbal cues don't suit.

e. *Disrespect*: There is never an excuse for disregard or rudeness. This sort of behaviour pushes people to the margins and makes their constructive engagement most unlikely. When they are marginalised, people do enough – but just enough.

In general, effective communication can be undermined in a pervasive way by an unconstructive mindset. This means there are habits which will be damaging and abrasive, but not necessarily conspicuous. These are habits of mind, which frustrate the properly fulfilling experience of organisational life for all people. They are private habits, and whilst they may surface in specific behaviours, they may arise in a less obvious permeating way and merely affect morale. Some of these unconstructive habits are

a. *Resistance to change an initial impression*: Some people enjoy a delusional confidence that they are always right. All people see things with perfect clarity from their own perspective. Few people are always right.

b. *The desire to defend ourselves from looking foolish*: The desire to save face is misguided. Everyone makes mistakes. This should be acknowledged and used as a basis for improvement.

c. *Withholding information and/or uncritically maintaining our opinion*: Only rarely will one person be in command of all the facts. Knowledge is not power. Only shared knowledge is powerful and only an open mind is truly constructive. More than just being open and curious, a constructive mind is sensitive to the fact that in evolving events, professional expertise is different from situational expertise.

d. *Blaming others when our message is misunderstood*: People who fail to acknowledge their mistakes will fail to learn from their mistakes. Such people fall short of greatness.

e. *Prejudice*: Prejudice can manifest in the halo effect, where we illogically overlook human failings and see only that which we want to see. The inability to see flaws in our own children might be a case in point. Equally, we can fall afoul of conspicuous and typical prejudices about things such as gender, race, ethnicity, religious doctrine or whatever – and let this get in the way of a clear-sighted perception.

f. *Complacency, fatigue and recklessness*: The inability to manage our obligations properly is amplified by fatigue or by a careless or incomplete appreciation of risks.

4.4 Effective Communication

Effective communication takes many forms. There is no one form of effective communication, no defined script. But effective communication is established on two broad principles:

1. Communicate respectfully and clearly, and check that your meaning is understood. This means: be explicit, ask precise questions and accept nothing less than direct, categorical answers. Be plain, be clear. But, be courteous. Respect is decisive. This is because the point or issue which is the substance of communication is, in some ways, less important than the means and method of communication. In other words, the method, style or manner of communication is critical to overcoming any potential misunderstanding.

2. Confirm that you have understood others correctly. This means: when there is no room for error, make sure that you have an absolute grasp of the precise and explicit meaning of the instruction or the information which you receive. In practice, this means that you might pose an absolutely unambiguous question, like this: 'Since there is no room for misunderstanding on this point, do I understand you correctly when I take your meaning to be X?' OR 'When you say X, do you mean X?'

4.5 Communication between Different Agencies

Plain and unambiguous communication is especially important when teams or groups from different agencies come together to accomplish a task.

We might imagine, for example, police, fire and ambulance services coming together at the scene of an accident, or police services, other security agencies and the military combining to deal with an incident. Very often, the military and emergency services cooperate in the response to civil emergencies. Or we might imagine industry, government and non-government organisations and agencies will come together in the response to an oil spill. Even at the level of the everyday, temporary teams are a derivative of the commercial practice of subcontracting.

Groups like this come together with a distinct purpose or goal, and with a distinct lifespan. These natural teams lack the formal structures and the deep personal understructures of trust and mutual confidence, yet they act with a practiced sureness, as if they had trained and worked together for years.

The spontaneous and immediate confidence which takes shape in groups like this has been called 'swift trust' [56, p. 167]. This term makes immediate and intuitive sense, depicting the professional and technical confidence, which connects diversely skilled people. But though groups like this depend on an elaborate body of knowledge and diverse skill, imperfect communication can mean individuals are not well placed to sort out just who knows what.

To minimise confusion or misjudgement, and to make the most of the interdependence and relationship, which energise multiagency groups, leaders must communicate very plainly. In these temporary groups, the special task of leadership is to foster the collaboration, the coordination and the communication, which will mitigate error and misunderstanding. The temporary, multiagency group is no place for competitive empire building. Constructive leaders will try to foster role clarity; they will enable reciprocal adaptation, the practical give and take of cooperative negotiation or exchange. In larger multiagency combinations, the use of liaison officers has been seen to be an effective way to resolve misunderstanding or difference between cooperating organisations, to enhance clarity and to make the most of the intuitive swift trust which unites people in a common enterprise [16].

In a nutshell, there are occasions in which there is no room for error or for discretion. On such occasions, assumptions can backfire and compound the problem. When there is no room for error, the person who speaks and the person who listens must be more than normally vigilant – making sure that he or she is understood and that he or she understands.

4.6 Effectiveness, Efficiency and Psychological Safety

The practical result of effective communication is in building a shared understanding and yoking together collective skills. In this way, effective communication renders individuals genuinely useful as part of a mutual whole. The phrase 'genuinely useful' points to the fact that risk tends to be minimised and missions tend to be accomplished more effectively and efficiently when people are on the same page.

The words *effective* and *efficient* can be overused. They are used here intentionally to mean

a. The term *effectively* is used to mean that a mission or task is accomplished totally, definitively and conclusively. Nothing is left undone. In other words, the details people needed to pay attention to have been understood, and attention has been paid to these details.

b. The term *efficiently* has been used to describe the proficient and practical use of resources. The idea is not cheap, but *value for money*. The idea is not effortless, but *competently*. Something is efficient when no money is wasted, when no effort is wasted, when every spillover and payback is captured.

Effectiveness and efficiency will be established upon a sense of psychological safety [21]. The idea of psychological safety describes the sort of organisational culture within which people have a shared confidence that the team is a safe place where people feel able to take an interpersonal risk. This means that the team is a place where people feel able to question decisions, to suggest innovations and to propose opportunities.

In this way, a climate of psychological safety – which arises from effective communication – will be the backdrop to constructive individual contribution, to initiative, to foresight and development and to effective and efficient accomplishment.

Five essential skills offer a foundation for effective communication. Leaders should be proficient in these skills and should foster these skills in others. The skills are

1. *Questioning*: Questioning is an antidote to misunderstanding. It is not insubordinate, though in power over cultures it is often mistaken for insubordination. Questioning is an obligation, especially when there is

a. Uncertainty about what should be happening

b. Uncertainty about what is happening

c. Discrepancy between what is happening and what should be happening

The absence of questioning indicates a problem. Silent colleagues are ineffective. At briefings, for example, questioning and proper exchange must be encouraged, not quashed.

2. *Assertion*: Assertion is a hard call – but all individuals have a responsibility to be assertive, especially when they are in a position to promote situation awareness, avoid error or avert disaster. Bad things happen when good people do nothing. Good people are respectful, and properly assertive in context [25, p. 615].

3. *Listening*: People listen with more than their ears. Listening involves sensitivity to the non-verbal cues which are essential to understanding. Listening also involves mental effort. To listen properly, put prejudice and bias in the background, listen for facts, create an environment which fosters communication, don't look for opportunities to be oppositional and seek constructive exchange.

4. *Focus*: Focus on facts and outcomes. Separate facts from the people who articulate them. Keep situation awareness; maintain a sense of perspective and the outcome for which everyone is working. Aim to address and to make clear WHAT is right, rather than WHO is right.

5. *Feedback*: After-action reports and post-incident analyses of lessons learned are formally structured critiques, which underline the importance of feedback. At every opportunity we should confirm our understanding and our sense of 'where to next'. This means that feedback is valuable when it follows careful analysis and takes a painstaking, structured form. And feedback is valuable when it is in the moment, given and sought in a constructive spirit. In this way, feedback is a dynamic element of the exchange in team problem solving.

In practice, these ideas inform what is often described as *support language*. Support language is a form of dialogue – a formally structured script, or a formal communication method or process – which emphasises the *professional* and deemphasises the *personal* nature of questions and answers.

A support language process might follow the format illustrated in Table 4.1, and summarised by the acronym RAISE.

What you see here are the formally structured steps in a communication process. The words which occupy the stages are not defined, as they will be

TABLE 4.1

Raising a Concern

R	A	I	S	E
Relay information.	Ask if they are aware, and seek clarification.	Indicate concern.	Offer a solution.	Emergency language: 'You must act now!' or a command such as 'Stop!'
	Inquiry		Concern	Emergency

found by people in the moment. But the stages are defined, and the intent of each stage is defined.

Importantly, people can commence the RAISE protocol from any stage. For example, people may find they are concerned, and realise that indicating concern is required – not asking questions. Similarly, observing an emergency, it would be perfectly sensible to simply command 'Stop!' without needing to pass through the preceding stages.

In the imaginary circumstance of an aircraft running low on fuel, here is a script which illustrates the support language idea. The idea is adapted from the work of Gary Klein and from Karl Weick [87]. In this imaginary incident, you are the co-pilot, and you are flying beside a pilot who ignores everything you say.

- a. *Gain attention*: Do this by addressing an individual directly – you might use a name, a nickname, a title or some other byword or term – but use it assertively: 'Bob!'/'Boss!'/'Captain!'
- b. *State your concern*: Do this by expressing your concern or analysis of the situation plainly and directly. For example, let's imagine a co-pilot who is concerned that an aircraft does not have sufficient fuel. The co-pilot says, 'I am concerned we do not have enough fuel....' *But, since you get nowhere, you go on. Now it is essential to*
- c. *Restate the problem as you see it*: Returning to the preceding example, the co-pilot does this by saying; 'We are showing fuel for only another 40 minutes...' *Again, you get nowhere. So, you must*
- d. *State a solution*: Returning to the preceding example, the co-pilot does this by saying, 'We need to divert to airport X immediately'. *Again, you get nowhere. So, you must*
- e. *Be assertive*: Do this by restating the problem and solution in clear and unambiguous language. For example, 'Captain, you must listen to me. We are running out of fuel; you need to divert to the airport now!' *Must is the definitive word here. But again, in this scenario you get nowhere. So, you must*
- f. *Exert or assume positive control*: Returning to the preceding example, the co-pilot takes control of the aircraft, and he says, 'I have control of the aircraft'.

The case of United Airlines Flight 232 is paradigmatic (see Box 4.2). This case illustrates the success of an aeroplane captain who led brilliantly in a time of enormous trial, because he recognised his own limitations and relied sensibly on others.

In a nutshell, the vignette in Box 4.2 illustrates the captain's willingness to accept help and suggestions from others and his call on others to contribute their own problem solving to manage the situation.

BOX 4.2   THE CASE OF UNITED AIRLINES FLIGHT 232

In 1989, United Airlines Flight 232 suffered an uncontained engine failure, which ruptured hydraulic lines, causing all three hydraulic systems to fail [26, pp. 254–255]. The crew was unable to exercise movement of any control surface. Aircraft altitude was managed by the application of asymmetric thrust.

The aircraft, a DC-10, was cruising at 37,000 feet en route to Philadelphia via Chicago. The aircraft was swinging through a gently banked near 90° turn on autopilot, to capture a new heading, when there was a bang and a lurch. The whole aircraft shuddered.

On the flight deck, instruments showed the number two (centre) engine had failed and was spooling down. Captain Haynes called for the engine shutdown checklist. In the process of carrying out the shutdown, Flight Engineer Dvorak was concerned that the hydraulic pressure and hydraulic fluid quantity gauges, for all three hydraulic systems, were falling.

Meanwhile, instead of straightening on a new heading, the aeroplane was continuing to swing to the right. It was also tending to nose down, in a descending turn. The first officer, William Records, disconnected the autopilot and attempted to straighten the aircraft with primary controls. There was no response.

Captain Haynes took the controls and equally found no response. He reduced thrust on the port side to restore wings level altitude. The captain also deployed the emergency air-driven generator, in order to power up the auxiliary hydraulic pump. However, hydraulic pressure could not be restored.

The aircraft was stabilised on a southerly heading, pitching gently fore and aft in a slow phugoid motion and rolling gently side to side in response to the pilots' management of asymmetric thrust.

Air traffic control gave the DC-10 a course for Sioux City Gateway Airport, six nautical miles south of the city, at an elevation of 1098 feet and on river flats adjoining the Missouri River.

Captain Haynes asked Senior Flight Attendant Janice Brown to report to the flight deck. He told her to secure the cabin, and to prepare the passengers for an emergency landing and evacuation.

With no manual back-up controls, and without hydraulic pressure, the only hope of controlling the DC-10 was by manipulating the thrust levels of the two remaining engines. This was a mammoth task for which none of the aircrew had been trained.

Recognising the gravity of the situation, a passenger, Dennis Fitch, identified himself as a DC-10 check and training pilot and volunteered his services to the flight deck.

Captain Haynes, the pilot in command, asked Fitch if he would take over the power levers, which Fitch did, kneeling on the floor between the two pilots and manipulating the No. 1 port engine lever with his left hand and the No. 3 starboard lever with his right. Manipulating asymmetric thrust, Fitch had to try to maintain a stable pitch altitude. To do this, he had to dampen out the low frequency phugoid oscillation, and the rolling of the airplane from side to side and the continuing trend to enter a descending turn to the right. This was vastly difficult.

The contribution of Check Captain Dennis Fitch enabled the operating crew to concentrate on critical individual tasks.

Captain Haynes obtained the approach speeds for a no flap no slat landing. He obtained the heading of runway 31, which was 9000 feet long. Haynes dumped fuel using the quick dump system. Fuel was jettisoned to the level of the automatic cut-off valves, leaving some 33,500 pounds (15,200 kg) on-board. Fitch was instructed to manipulate the power levers so as to maintain a 10–15 degree turn to the left. This was an almost superhuman task – to which Fitch devoted his complete attention.

Twenty-one nautical miles northeast of the airport, the airplane was about four minutes from landing, but unable to make course adjustments necessary for an approach to runway 31. The shorter runway 22, which was only 6600 metres, was selected instead.

Fitch used all of his 23,000 hours experience to work the power levers in order to maintain a descent profile, which smoothed phugoid oscillation and the roll motion of the aircraft. He needed to maintain a speed of between 200 and 185 knots, bring the airplane over the runway threshold at about 100 feet, and on a relatively straight heading.

The aircraft struck the ground with the starboard wingtip, just short of the runway threshold and just to the left of the runway centerline.

The aircraft broke up on impact and turned over, wreckage skidding for more than a kilometre.

Of the 296 occupants of the aircraft, 110 passengers and one flight attendant perished. This loss would have been much worse, if it was not for the proactive management of a dire situation.

The loss of all three hydraulic systems resulted in the crew being deprived of all moveable surface control. Additionally, the crew combated airframe vibration and in-flight oscillation, which made controlled flight impossible.

The case demonstrates the importance of collaborative *power with* cultures and the capacity, which comes when individuals combine and cooperate – even in highly tense and pressurised moments. The significance of these remarks is underlined by many cases in which a difficult institutional culture and perverse, uncooperative individual behaviour have hastened unfolding disaster.

Less obviously – but very powerfully – the vignette reveals the fruits of institutional leadership, which inculcated habits of self-awareness, constructive communication and integrated teamwork. The next chapter discusses the ways in which these attributes come together and support situation awareness.

5

Leadership and Situation Awareness

5.1 Introduction

Chapter 4 spoke about the idea of effective communication, which was seen to be an important *power with* leadership. The practical upshot of effective communication was seen to lie in generally effective and efficient teams, groups within which people felt psychologically safe and, for this reason, able to participate, question and innovate.

This chapter illuminates the idea and explains the critical concept of shared situation awareness. As well, the chapter demonstrates how leaders can create the background conditions to support shared situation awareness.

5.2 Situation Awareness

Responsible judgements, decision and action depend on accurate situation awareness, the single most important factor in improving mission effectiveness, which is understood as: 'The accurate perception of the factors and conditions that affect (events) during a defined period of time. In simplest terms, it is knowing what is going on around you – a concept (which) embrace(s) the need to think ahead' [63, p. 256].

Another applicable definition is: 'Situation awareness is the perception of the elements in the environment within a volume of time and space, the comprehension of their meaning and the projection of their status in the near future' [63, p. 257]. Taking this point, Endsley [24, pp. 9–10] identifies three degrees of failure to accurately perceive and interpret situations:

a. Level One – Failure to perceive the situation correctly

b. Level Two – Failure to comprehend the situation

c. Level Three – Failure to comprehend the situation into the future [25, p. 609; 63, p. 265]

The Endsley taxonomy illustrates how accurate situation awareness depends on perception and understanding. But it does not speak to the complication which arises in complex interdependent systems, where a distributed (or shared) situation awareness is critical.

The ambition of distributed situation awareness is that a compatible understanding of events is shared across a team. The reason for this is: in highly technical, or highly regulated environments, time pressures will likely overwhelm the individual – but cohesive teams, with the capacity to share workload, will likely survive.

This ambition of distributed situation awareness gestures to the tendency for different teams and different team members to view and use the same information very differently [80]. To be explicit, there is a natural variation in situation awareness *within* teams and *between* teams, which is amplified when teams from different organisations or agencies cooperate. For example, the police, ambulance and fire services often cooperate with each other, and teams from each of these agencies cooperate with the citizens.

At each point of intersection – when one person unravels real-world events, when one person cooperates with another, or when one team cooperates with another – there is the potential for misunderstanding. Potential for misunderstanding is amplified wherever there is a language difference: wherever words are used to mean different things. For example, a citizen describing a fire as 'really bad' might be expected to have a different view of the world from that of the professional fire fighter who sees 'really bad' fires quite often. But perhaps not. Perhaps the description offered by the citizen matches the perception of the fire fighter. The point is: nothing can be taken for granted.

The potential for misunderstanding is also amplified over time because people become accustomed to their colleagues and to the pattern and routine of their operating environment. The phrase 'becoming accustomed' describes the comfortable familiarity, the normalised deviance or the practical drift which precedes error and accident.

An example of the way in which local practice in a distributed system may drift to become disconnected is offered by the destruction of two US Army Black Hawk helicopters over northern Iraq by two US Air Force fighter aircraft. Analysis [79] concluded that the pressures of local operational practice induced a shift to locally efficient but globally inconsistent procedures. This practical drift led to failure of coordination at broader levels.

5.3 Generating Situation Awareness

So we ask, how can a distributed situation awareness be both tailored enough to meet different needs, and robust enough to be resistant to coordination failure?

The critical element is communication. Leaders need to be good communicators, and they need to teach and train and encourage all of those with whom they deal to be good communicators as well (Box 5.1).

The message of situation awareness is in the meaning of perception and interpretation. Effective communication will aim to convey and to foster a grasp of all factors or elements within a defined context, and the capacity to think ahead and to anticipate. Doing this, an individual will be asking the 'what next?' and 'so-what?' questions. 'What might happen next, what will this mean to me, how might I respond most constructively?'

When we say effective communication will seek to foster a grasp of factors, and the capacity to think ahead and to anticipate, we are talking about the way that effective leadership gives expression to the *power with* idea. Rather than telling others what to perceive and how to understand the world around them, leaders enable and encourage people to build distributed situation awareness.

How do leaders do this? How do leaders secure the background conditions which are critical to distributed situation awareness? One of the key things is for leaders to develop, among all people, the preparedness to challenge the understanding and judgement of one's superiors, one's juniors and one's own. Leaders will foster this preparedness by their example, and by the climate and culture they build.

Preparedness to challenge does not mean challenge tactlessly or merely for the sake of challenging. The preparedness to challenge is a professional skill and requirement, which entails continual vigilance and alertness in order that perception and conclusions are subject to unfailing scrutiny. Nothing must ever be taken for granted; no observation and no conclusion must ever be accepted on face value. *Everything must be the subject of intellectual scrutiny.*

In a nutshell: in the moment of crisis, grave danger or emergency, purposeful action can be taken with confidence and assurance, as no one will be wondering whether the situation was like this, or like that. Confidence and

---

BOX 5.1 THE IMPORTANCE OF TESTING ASSUMPTIONS

Gentlemen, if we are all in agreement on the decision – then I propose we postpone further discussion of this matter until our next meeting to give ourselves time to develop disagreement and perhaps gain some understanding of what the decision is all about.

Alfred Sloan
CEO, GM, 1923–1956 [14]

assurance arise from shared situation awareness. Additionally, any verbal challenge, and certainly any public challenge, requires

a. Tact, and appropriate tone

b. Good purpose. This means that a challenge needs to be relevant and important – a matter of significance, not of mere preference or personal advancement

c. Good, solid, rational grounds based on careful observation and scrutiny. Leaders want to avoid situations like 'the boy who cried wolf'. It is important to avoid an environment which is so liberal that challenges arise with irritating frequency on the basis of cursory scrutiny. The leader will want challenge only on the basis of careful professional scrutiny and genuine concern.

5.4 Leading and Cooperating

The best leaders and followers cooperate to fulfill duties, even when duties are burdensome or not a matter of preference. The best leaders and followers recognise they are together in a strict seniority system and a dynamic environment [63, p. 281]. The logic is illustrated by comments made during the Royal Commission investigating the 28 November 1979 crash of Air New Zealand DC-10 (Flight 901) into Mt. Erebus, Ross Island, Antarctica:

> It is the inherent responsibility of every crew member, if he (or she) is unsure, unhappy, or whatever, to question the pilot in command as to the nature of his (or her) concern. Indeed, it would not be going too far to say that, if a pilot in command were to create an atmosphere whereby one of their crew members would be hesitant to comment on any action then *he would be failing in his duty as pilot in command*. [63, p. 285]

This is not to say that people in responsible positions will discuss and collaborate on every decision. To do so is unnecessary, and it would become very tedious. The point is this:

a. No subordinate should hesitate to ask the reason or rationale for decisions he or she does not understand, or which are not routine.

b. People in senior positions should make it possible for people in subordinate positions to pose *professional* questions.

c. Seniors, like captains in command of aircraft, should brief regarding the contingencies and individual responsibilities before crucial moments.

d. Seniors in particular, and people in general, should take every chance to capture the lessons learned from significant incidents. Debriefing is critical – people won't learn and improve unless opportunities for learning and improvement are seized.

These ideas have been applied and researched in surgical teams involved in minimally invasive cardiac surgery. Edmondson [22] investigated the adoption of new technological procedures and found the most successful teams had leaders who emphasised the need to pool experiences and learn collectively. These teams demonstrated 'a paradigm shift in how we do surgery, (because) the whole model of surgeons barking orders down from on high is gone'. The research found

a. The ability of the surgeon to allow himself to become a partner, not a dictator, is critical.

b. (More approachable surgeons) led others to report that they felt respected for their particular abilities and contributions.

c. (Surgeons) encouraged speaking up by communicating a sense of humility, such as by noting (their) own fallibility or limits. One surgeon commented, 'The realisation is that you cannot manipulate all the variables in the procedure yourself'.

d. This observation both conveys humility or limits *and* points to a clear need for others' input. Such actions served to deemphasise the surgeons' power, reducing this potential barrier to speaking up, and reducing the risk of catastrophic error by empowering junior team members [22, p. 1440].*

However, whilst dominance is not constructive, neither is passivity. Excessively congenial people may be seen as passive, or perhaps reflexive. Excessive congeniality is not helpful and does not inspire confidence in a team. Worse, passivity leads to risk when people *assume* they share the same understanding without actively checking this out. Under these conditions passivity leads to false consensus and collective ignorance [27]. The balance is fine. The balance is not one that can be properly defined by words. Essentially, one must lead, and one must cooperate or, in a term favoured by this book: *collaborate*. The idea is that leaders exercise power with the team members on whom they depend. They do not exercise all the power and they do not fail to exercise power. Leaders exercise responsibly – they are good collaborators. In a phrase which suggests this idea: 'Leaders have a loose rein, but a firm grip'. This means they exercise just the right amount of 'adult supervision'; they are not overbearing or overwhelming, and neither are they irresponsibly hesitant or unsure.

---

* Slightly abridged.

People need to be confident and competent, not overbearing and not over-confident. There does need to be a sense of humility – perhaps modesty, or circumspection or unpretentiousness are perhaps better words, because humility may suggest an uninspiring sense of fallibility, which can induce others to lose confidence and to panic when things go poorly.

The best leaders are confident, not arrogant; they are collaborative, yet decisive, and sufficiently confident to take responsibility for the decisions they take. Another way of making this point is to say that responsible people will make the best decision possible, using all the resources available, and *in the time available* [63]. A good leader develops, rather than dominates, a good team.

Team development is much more than fostering friendliness among people. A properly developed team communicates well and builds shared situation awareness. At one level, situation awareness is purely individual, but it is also a team concern because the complexity of interconnected events, in highly technical and time-pressured environments, will likely overwhelm the individual – whilst cohesive teams will likely cope.

Bridging the gap between an entirely individual perspective and a collective or shared appreciation of events, effective emergency management team leaders engage in three types of coaching behaviour:

a. Boundary riding: assisting team members to be in temporal alignment with one another's needs;

b. Boundary spanning: assisting team members to be integrating their internal activities within the team

c. Boundary crossing: assisting team members to work with others outside the team [64]

5.5 Situation Awareness: In a Nutshell

This book claims leaders must work at their own situation awareness, and at the same time generate the background conditions which are critical to distributed situated awareness in a team – or between teams. The concept is complex. A complete analysis of the psychology is beyond the scope of this text.

In a nutshell, in order to develop situation awareness in oneself, and distributed awareness in a team and between teams, leaders must do these things:

a. Monitor developments

b. Anticipate required actions

c. Ask the right questions

d. Test assumptions; confirm understanding

e. Monitor workload distribution and fellow crew members

f. Report fatigue, stress and overload in oneself and in others

A loss of situation awareness may be indicated by the following factors:

a. *Ambiguity*: where a person is uncertain of his or her perceptions, more than one interpretation is possible and distinction between possible alternative readings of events is unclear.

b. *Distraction*: where attention is not properly focused on key issues or events and easily diverted by side issues.

c. *Fixation*: where attention is focused closely on one issue or event, to the exclusion of a wider perspective.

d. *Overload*: where too much is happening and there is not enough time to manage all events and issues properly.

e. *Complacency*: routine, established habits and conventions mask wider awareness of broader issues and looming threats or challenges. A failure to ask the 'so what?' and 'what if?' questions.

f. *Deviation*: abandonment of standard operating procedures without good reason – for preference or convenience – and for no other more compelling reason.

g. *Avoidance*: a failure to be vigilant. Putting one's head in the sand and pretending that everything is fine. Failing to resolve conflict or discrepancy.

5.5.1 Five Ways to Improve Situation Awareness

There are a number of things a leader can do to develop and maintain shared situation awareness. These include

a. Plan ahead and predetermine specific roles for your team. Assign responsibilities for handling problems and for unexpected distractions.

b. Know your people and all the other resources available to you – know who is well qualified to assist and how. Actively source input and the perspectives and opinions of others.

c. Avoid fixating on one specific problem or one part of a problem. Direct your attention systematically to all the decisions – and the elements of these decisions – for which you are responsible. Repeat this pattern over and over.

d. Question by considering 'what if?' and 'so what?' scenarios. Anticipate; make a deliberate disciplined effort to avoid being taken by surprise.

e. Speak up if you detect signs that situation awareness in your team is degrading [24].

5.6 An Important Caveat

Even an autocrat who adheres strictly to these methods will fail. Aggressive questions will not generate constructive answers. An aggressive question will provoke nothing more than staff answers – compliant and essentially unconstructive responses. Aggression can be communicated in countless subtle ways. Tone, expression, gesture, posture, and language can all communicate aggression or unhelpful belligerence. Skilled leaders will be sensitive to the context, which is hugely significant.

6

Leadership and Decision Making

I shall show you a Thinker: A will undisappointed, evils avoided, powers daily exercised, careful resolution, unerring decision.

Epictetus

6.1 Introduction

Chapter 5 spoke about the idea of situation awareness, and about situation awareness being shared or distributed within a team or between teams. The practical upshot of situation awareness was seen to be a collective understanding and consciousness, which enabled uncertainty to be minimised. In consequence, risk was understood to be managed and, in time of crisis, decisions were understood to be made with confidence and assurance.

This chapter examines the idea of leadership and decision making. This is an important topic, as action is best when it is based on rational decision. And decisions are best when they are based on a sound appreciation of the general structure, pattern and the dynamics of decision making.

6.2 Elements of Decision Making

In broad terms, making a decision involves

a. Choice among alternatives
b. Accurate assessment of the specific nature of a problem
c. Risk assessment, which may or may not be explicit, but which is inherent in every single decision
d. Action
e. Reassessment and reaction

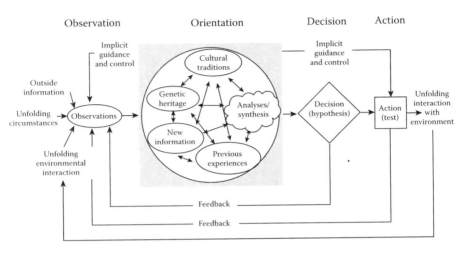

FIGURE 6.1
John Boyd's OODA Loop in flowchart format. (Source: McIntosh, S. E., The Wingman-Philosopher of MiG Alley: John Boyd and the Ooda Loop. *Air Power History*, 2011. 58(4):24–33.)

The way these elements might be addressed and weighed, each against the others, is diagrammed in the OODA Loop, which is shown in Figure 6.1. The OODA Loop – Observe, Orient, Decide and then Act – is a model, devised by Colonel John Boyd of the United States Air Force [55]. The OODA Loop acknowledges that to make timely decisions, people must be able to form mental concepts of observed reality, and be able to change these concepts as reality itself appears to change [77].

The phases of the OODA Loop are

a. Observe: this means utilise all possible means to gain awareness.

b. Orient: this means make sense of what is learned.

c. Decide: this might be a rule-based decision or a creative/naturalistic decision (these are discussed later).

d. Act: build a course of action and give effect to this.

The orientation phase is the *schwerpunkt* or decisive point of the model. Schwerpunkt, a German word, describes a decisive point, or focal point or the *centre of gravity*. The schwerpunkt is the heart of a problem, or a point of emphasis. It is where effort should be concentrated, so a decisive result might be achieved.

In John Boyd's model, the orientation phase is the schwerpunkt, the decisive point, because orientation presumes understanding. Orientation takes it that people will have a grasp of the factors, which must be addressed, and the *type of decision* which must be made. This logic coincides with the ideas of bounded rationality articulated by Herbert Simon in 1956. Simon argued that 'a great deal can be learned about rational decision making by taking into account, at the outset…the environments to which it must adapt' [76, p. 129]. The decisive point is that rationality is relevant to context. This means: decisions which are rational for one person at one time might not be so sensible in another situation or for another person.

Ideas of risk assessment and rational choice are illuminated by a brief survey of decision theory.

6.3 Decision Theory

Decision models describe, not what decision makers actually do, but what they should do. Most models are linear and describe ideal decisions in ideal circumstances. They are typically informed by mathematical calculations of subjective probability [30].

Linear decision models which are useful but limited reflect the ideas of Brim that a rational decision process should follow the sequence:

a. Identify the problem.

b. Obtain essential information.

c. Devise possible solutions.

d. Evaluate solutions.

e. Devise a strategy for performance.

f. Enact that strategy, [8, p. 9].

The acronym D-E-C-I-D-E-R encapsulates this logic.

- **Determine** the problem.
- **Evaluate** the scope of the problem.
- **Consider** options to either mitigate or resolve the problem.
- **Identify** the most appropriate option.
- **Determine** an action plan and **DO** this.
- **Evaluate** the effectiveness of actions.
- **Re-plan, Re-brief** and **Re-act**, where necessary.

The value of linear models like this is in the *demonstration* of rational decision strategy. The model is useful because it is *illustrative*, not because it is definitive. The model should be seen as offering a suggestion, offering an instance of the rational steps a person might follow in the ideal world, the linear model – like all models – offers possibilities, not definitive answers. This means: models or rubrics like this are not scripts to be followed rigorously.

In a nutshell, no model and no rubric translates seamlessly into real life since, as Witte established, people cannot 'gather information without in some way simultaneously developing alternatives...evaluating these alternatives immediately (and thus making) a decision' [89].

Non-linear decisions tend to be the actual practice in real life. In life, 'rules [or rubrics] are never a substitute for good judgement' [63, p. 277]. This means decisions, which may well be informed by rules, should not be enslaved by rules – or by the rigorous steps of a decision model. When a rule or model does not fit a situation, then responsible judgement must apply. Judgements will be informed by rules and rubrics, and also by experience, practice, by the information available, by the practical demand for flexibility and by the need for urgency.

Real life is *naturalistic* or *event-driven*. In life, people use experience. In real life, where decisions are event driven, people rely on practical wisdom – the knowledge gleaned from what they have observed, encountered and endured.

Experience means people are able to recognise, in the particular events or circumstances they face, similarity or correspondence with past events. Relying on this experience, people come to make what are known as pattern-matched decisions. The pattern-matched decision, known by an alternative form as a recognition-primed decision, is nonlinear. The pattern-matched decision derives from the mnemonic of lived experience. The present cue sparks the awareness of previous and similar situations, of formerly successful courses of action and earlier outcomes. Such a decision is an efficient way of responding in times of stress or time pressure. Informed by experience, analysis of a situation is faster and response is faster. If an initial evaluation is wrong, then courses of action are adapted.

The idea of the pattern matched or recognition-primed decision arose from the study of fire fighters by research psychologist Gary Klein. Klein found that fire officers, robbed by time of the ability to analyse options, adapted prior experience to present situations [47,48]. Judith Orasanu from the NASA-Ames Research Center explains that when a problem is identified as matching a particular type, the expert retrieves a response from memory. The response is evaluated for adequacy. If found wanting, a further response is selected. Alternatives are evaluated in sequence, not concurrently [61]. But alternatives do not tend to be evaluated each against the others. And, expert practitioners who are accustomed to this mode of thinking are less likely to generate or consider multiple alternatives [30].

The same phenomenon is apparent when doctors make diagnoses – symptoms are considered and a diagnosis made according to what fits. Similarly, novices become experts; developing skills and understanding through education and experiences; eventually combining their knowledge of why with their knowledge of how [19] (Box 6.1).

BOX 6.1   WHEN PERSONAL EXPERIENCE WON'T HELP

On September 11, 2001 Arlington County Deputy Fire Chief James Schwarz was faced with managing a fire and rescue operation resulting from a plane crashed into the Pentagon building. Although he had managed many fire events this was unusual and not one he had stored memories to call upon. Instead he recalled the Alfred P. Murrah building which was destroyed by a bomb on 19th April 1995. Deputy Fire Chief Schwartz made contact with Gary Marrs who was the Fire Chief on that day in Oklahoma City. He knew The Chief would have the stored patterns required to give him the advice he needed.

D. Parsons by James Schwarz at the Leadership in Crises workshop
Harvard Kennedy School, 1–6 April 2013

Methodical systems of pattern matching work well when a response must align with specific decision rules. Newell and Simon offer an example, explaining chess, not as a series of random moves, but a series of responses – each matched to the system of decision-making rules [59]. Pattern matching works less well in unpredictable and unfamiliar contexts, which call for spontaneity or creativity [58]. Ambiguous contexts call for more creative and innovative solutions. There is, in such circumstances, significant potential for decision makers to draw on the skills and talents of the team.

Hence, decision-making research has 'brought a renewed focus to studying teams in context' [51, p. 341]. There is a need to understand the way in which people work together in order to make the best decision possible, to minimise error and to learn from error.

6.4 The Collaborative Approach

The collaborative ideas, which are central in this book, provide a framework within which individuals can contribute to creative and conscientious decisions. The collaborative model plays down the tradition of the dominant and foolproof leader. The collaborative approach emphasises the obligation of *people as responsible individuals* within a team to identify, trap and correct mistakes rather than play along passively as mistake turns into disaster.

The collaborative approach will not eliminate error. But the collaborative approach will enrich decision-making habits and help people and their organisations to minimise and to manage error (Box 6.2).

BOX 6.2   DISTRIBUTED DECISION MAKING IN A POWER WITH MODEL

At the time of writing, the NSW Rural Fire Service has a situational awareness computer program called ICON. ICON enables all users to see live operational information such as operational maps, operational logs, thermal image scans of fire boundaries, aircraft movements and fire weather forecasts. Numerous government agencies with a role to play in bushfires have been given access to view ICON. This empowers dozens of agencies with information to make operational information available to inform their own decision making during a bushfire emergency without centralised direction.

NSW Rural Fire Service

The collaborative approach helps in two main ways. First, the collaborative *power with* approach is an antidote to the domineering *power over* styles, which make it preferable (psychologically safe) for junior people to remain disengaged and uninvolved. The danger is that junior people, who seek to be psychologically safe, find this safety in their disengagement. When they are uninvolved junior people do not alert senior people to oversight or error, or to risk. No accident may ensue. But, since junior people are discouraged (by the unpleasant effects of dominant power over behaviours) from speaking up, the risk of accident is elevated.

Second, in a collaborative climate, it is easier for people to own mistakes, and thus to learn from the errors, which are inevitable given the natural limitations of human performance and the function of complex systems. This is important because one of the main drivers of excellence is the opportunity for people to speak plainly: to air views, to talk about errors and uncertainties, and to learn from the discussion. When this sort of constructive openness is not fostered, and people are led to believe their views are not respected, the outcome is a lack of free-flowing information, which is the most valuable currency of any organisation (Box 6.3).

BOX 6.3   AUTHORITY GRADIENTS IN THE US ARMY

Underlining the organisational corrosion which follows from excessively apparent and steep power gradients, the US Army 2012 Annual Report observed (p. 8) that 'too many junior officers believe that they will be punished if they offer senior leaders opinions judged as "too candid". Moreover, a majority of the Army's young leaders are convinced that a single mistake can end a career' [83].

The United States Army Report excerpted in Box 6.3 [83] makes a case for leaders everywhere. People must be aware of and sensitive to excessively prominent power gradients in their workplace. More than uncomfortable or unpleasant, over-steep power gradients are noxious, particularly because they can make it impossible for subordinates to 'fess up'. When errors or shortcomings are covered up the result is an escalation of risk.

These ideas are further informed by the idea of uncertainty avoidance, which coincides with the idea of groupthink, and refers to the need for rule-governed behaviour and clearly defined procedures. An organisation built around hierarchies and standard operating procedures – like an emergency response organisation – would seem to be home to a high uncertainty avoidance culture.

When high uncertainty avoidance is combined with high power distance, the upshot may be a culture of inflexible, unresponsive behaviours, dependent upon automated systems and an unwillingness to take personal responsibility or to make personally responsible judgements [35, p. 66; 75, p. 325]. Leaders must guard against this sort of suppressive stagnation; they must be prepared to make the hard right call over the easy option.

6.5 Collaborative Leadership and Decision Making under Stress

The benefits of collaborative leadership were conspicuous among the findings of a seven-year research program known as TADMUS – tactical decision making under stress [10]. Sponsored by the United States Office of Naval Research, the TADMUS program aimed to develop an understanding of critical thinking in stressful teamwork contexts, so that people might be trained to better deal with these situations.

TADMUS was initiated after the 3 July 1988 tragedy involving the USS *Vincennes*. This warship, a Ticonderoga cruiser, shot down Iranian Air Flight 655 over the Persian Gulf, killing 290 civilians. In this event, stress, task fixation and the unconscious distortion of data all played their part. Decision making was flawed and tragedy ensued, largely because decision making was not properly understood.

Acknowledging several different research paradigms, TADMUS interrogated and explained the typically complex relationship between task-work and teamwork [11, p. 25; 78, p. 63]. This is to say, TADMUS investigated the relationship between relatively mechanical, quantitative and foreseeable decision-making tasks and qualitative mastery, which enables different individuals to combine as an effective team, especially when that team was making decisions under stress.

The underpinning premise of TADMUS is that proficient decision makers are metacognitively skilled. This means that they are able to

- Identify evidence–conclusion relationships within an evolving situation. In other words, they could monitor the validity of a conclusion even as events were unfolding. This means they could judge whether the conclusion continues to hold, or is it no longer sound based on evidence of the unfolding event.
- Critique an identified problem within the situation to identify weaknesses. In other words, proficient decision makers could identify incompleteness, unreliability or conflict in the evidence as the evidence unfolded.
- Correct so that they can better respond to these problems. Correcting may require collecting additional data or shifting attention.

TADMUS is an established value of training teamwork skills and training people to develop shared situation awareness about an evolving event.

A related and valuable idea comes from Karl Weick [86]: the idea of *sensemaking*.

Weick established that when people focus excessively on the decisions which they make themselves, and which other people make, then the decision comes to be attached to ego.

When a decision is identified with ego, there is an unhelpful tendency to become defensive. This means that questioning takes on the sense of a personal hostility. Hence, in response to questions, people come to be self-justifying and protective. Weick [86] suggests it is more constructive to reframe the process of decision making as the process of sensemaking. When engaging in sensemaking it is easier to engage with other people, easier to update the information we have and easier to test the plausibility of the meaning we are making of an evolving situation. Sensemaking is an act of discovery; it is a collaborative and shared evolution – there is nothing about sensemaking which suggests leaders have to defend a decision or a determination. Sensemaking allows for correction, and it suits complex and rapidly unfolding situations. Significantly, it is the rapidly unfolding and complex situation which has traditionally been home to command and control direction. But, as Tannenbaum, Smith-Jentsch and Behson [82] observe: when situations are complex and unfolding rapidly, the traditional command and control approach is unlikely to have a constructive effect. In rapidly unfolding and complex situations, what will matter is flexible and adaptive coordination among team members. This sort of performance will require collaborative leadership – not directive command and control [82, p. 249].

In a nutshell, when leaders foster a culture of sensemaking, they foster the capacity to be flexible, to be responsive and to be cooperative. In such an environment, questions come to be appreciated as the antidote to uncertainty, and changing your mind comes to be seen as making a correction and not as evidence of indecisiveness.

6.6 Lessons Learned: Training for Good Decisions

Decision theory, like any theory, is just a theory. To be properly useful, theories must be coupled with realistic training and be responsive to the dynamic experience of operations. To learn from practice, after-action reviews must be established as a part of routine business. The aim of the after-action review must be to help people learn from shared experiences. This means after-action reviews will not be written from the top down in a sort of sanitised bureaucratic form of propaganda. After-action reviews must capture the most relevant experience so as everyone can learn, from the top down and from the bottom up.*

Following the scrutiny of real-life experience, training programs will not be unvarying and always the same. Rather, as training captures experience, training will evolve and change. This is because training should be focused on success in the field, not on compliance with the bureaucracy: training should be concerned with more than the administrative particulars of course accreditation.

Training must be demanding, realistic and connected to the real world. Training must enable people to capture the hard lessons of real-world experience. Without relevant, comprehensive and practical insight, training is often the source of rote-learned mistakes. But, just as the training environment must reflect the hard-won lessons of the school of hard knocks, so must practitioners be mindful of new thought. Without an awareness of fresh and evolving ideas, practitioners just get better and better at repeating the same old missteps.

In short, the real world is not a static unsurprising place. Any leader, sticking stubbornly to standard procedures without a mind to real life, is just pig-headed and mulish. Leaders must know established procedures. But a leader also needs the personal courage and the professional judgement to deviate when rules or standard operating procedures no longer fit the circumstances.

This is personal courage, which will be tested very much by the pressures of public scrutiny and the rigours of public inquiry. No leader could be unmindful: decisions made in the heat of the moment will be sized up in the court of public opinion, and possibly interrogated by a Board of Inquiry.

---

* This simple thing is hardly ever done. Consider, for example, the building evacuation. After the annual drill – which is inevitably inconvenient and a little shambolic – the floor wardens and supervisors will have a debrief. Amongst themselves they will decide that it all went very well. But they never ask for input from the hundreds of their innocent victims, the ordinary working people who are mistaken for puppets with nothing useful to say.

But a leader will do the right thing. This is one attribute that really counts – and it counts in crisis and in calm. Do the right thing. Be steadfast and straight up. When people have confidence that the leader is absolutely free from self-interest and cowardice, that the leader is more than a bureaucratic rule-follower, they will offer trust and confidence and respect.

7

Error Management

The preparedness to learn new things is foundational to sustainable organisational performance over time and in ever-changing circumstances.

J. S. Carroll and B. Fahlbruch
12, p. 3

7.1 Introduction

Chapter 6 examined the idea of leadership and decision making. The chapter explained decisions are best when based on a sound appreciation of decision making.

The present chapter explores the idea of error and the idea of error management. In today's technological society, and in large-scale automated systems, the consequences of error are infinitely greater than ever before [34]. The importance of this chapter is amplified by its universal relevance, as error does not respect age, rank or gender. The best people have made some of the worst mistakes. No one is immune from making an error [70].

There is an apt aviation human factors saying that both sides of the cockpit burn equally well, meaning that people who remain passive or silent find themselves impacted equally by the errors of their peers or leaders. All team members have an obligation to identify and act on mishaps or error trajectories before they become accidents.

Leaders have an obligation to foster the sort of climate where people find it easy to speak up. Leaders build a culture within which people find it easy to speak up and it is easy to 'fess up', and within which covering up is unlikely.

7.2 Terms: Error and Violation

In this chapter, terms are used in a very specific way. The term *error* is used to describe the sort of miscalculation or misjudgement, which it is normal for humans to make. The term error, then, describes human error, a mistake in belief or judgement – which very often translates to a mistake in action. Importantly, an error is not intentional. There is no wickedness or deliberate intention to do wrong, or to deviate from established or correct procedures.

In this chapter, the term *violation* contrasts with the term *error*. The term violation describes an intentional deviation from established rules, conventions or practice, and from correct methods or modes of operation. A violation is a deliberate infringement or transgression, a deliberate breach of an expectation, a trust or a directive.

Drawing these ideas together: error events are thus human mistakes, lapses or slips. Violations are characterised by deliberation and by the intention to breach. A violation may not necessarily be wicked, but it will always be deliberate. An example of a deliberate act which is not necessarily wicked would be the case of the person who deliberately breaks a speed limit to get a person to hospital. This chapter does not discuss those cases.

In considering errors, this chapter acknowledges that the causes and the consequences of similar errors can be quite different [34].

7.3 Errors Are Consequences

This chapter claims errors are consequences, but not causes, that is, errors do not occur as an isolated glitch in someone else's mind. Errors are shaped by circumstances: by the task, the tools and equipment and the workplace in general [70, p. 10]. Everyone has a responsibility to avoid error and to correct errors.

Understanding errors, their own susceptibility and the vulnerability of others, leaders will have a critical empathy and an enlarged awareness of the part they might play to minimise error and to learn from it.

7.4 Four Broad Error Types

Human error may be analysed broadly on four levels:

1. Unsafe acts of operators
2. Preconditions for unsafe acts
3. Unsafe supervision
4. Organisational influences

Such a framework (known as a human factors accident classification system) asks us to ask more sophisticated questions than, 'Where did the operator go wrong?'

Awareness of error types enables a sharper understanding of the reasons error occurred in the first place. The crucial point is that 'errors are viewed as consequences of system failures, and/or symptoms of deeper systemic problems; not simply the fault of the employee working at the "pointy end"' [65]. Having such a framework or paradigm (designed to be specific and tailored to the environment) enables the interleaved complex of human factor/human error to be broken into something like its constituent parts and understood [74,88].

7.5 Error in the Technical and Automated Environment

Error takes on a new significance in the modern, highly technical and increasingly automated world. This is because automation does not eliminate error. Neither does automation reduce the need for effective communication among team members, and for diligent and energetic leadership. In fact, automation actually 'requires a greater level of communication among crew members to ensure that everyone understands what is taking place' [35, p. 65]. This is because, unlike those in manual systems where everything is visible, inputs to computers are not readily apparent. Errors are therefore very easily disguised, and may be unobserved until the consequences are unavoidable.

7.6 Human Factors Underpinning Error

Errors may derive from untold human factors operating in combination or alone [25]. Conspicuously, *fatigue* is connected to performance impairment and error. Research demonstrates restriction of sleep results in significant deficit in cognitive performance on all tasks [42,67,84].

Error is also conspicuously attributable to work overload. This is because work overload can be the foundation of the perceptual or cognitive anopsia known as *inattentional blindness,* which is the failure of people, when they are focused on specific details, to see general context. In simple terms, inattentional blindness is the inability of people to see the forest for the trees.

The phenomenon of inattentional blindness is demonstrated by the basketball game and gorilla experiment. People were asked to watch a game between a team in light shirts and a team in dark shirts. They were asked to count passes made by the team in light shirts. A person in a gorilla suit

is not a concern. This means we do not find it difficult to apply techniques which, when we come to pull them apart and to analyse them, can be elaborate and terrifically convoluted.

We use heuristic techniques the way most people use motorcars. That is to say, with complete confidence the car will work, with a barely marginal understanding of the engineering on which the working depends.

But heuristics are fallible. Illusions demonstrate this fact. Illusions play on the heuristic tools we take for granted, and they continue to 'work', even when we know they are tricks.

For example, in the illusion shown in Figure 7.1 [73] we allow our eye to tell us that check A is darker than check B. We are tricked by a phenomenon called local contrast [68, pp. 7–8]. This means we interpret a check that is lighter than its neighbours as lighter than average. The light check in shadow B is surrounded by darker checks. Thus, the physically dark check B is light *compared to its neighbours*. The 'dark' checks outside the shadow are lighter than the 'dark' checks inside the shadow. But, surrounded by lighter checks, they appear dark by comparison. The contrast strips attest to this description.

The risk of deception which arises from the fallible unconscious perception tools we use all the time is amplified by the human tendency toward overconfidence, by the reluctance of junior people to challenge senior people and by the disinclination of most people to admit mistakes. In very simple terms, the *psychological need to protect our own self-concept* tends like this: if I admit a mistake, or a misjudgement, then I will look foolish/weak/incompetent and so on and so forth. Freud offers an example:

> I was going through my medical engagement book so that I could send out my accounts. Under the month of June I came across the name 'M-I' but could not recall whom it belonged to. My bewilderment grew when I turned the pages and discovered that I treated the case in a sanatorium and made daily visits over a period of weeks…. Finally the record of the fees I had received brought back to me the facts…M-I was a

FIGURE 7.1
Adelson's checker shadow illusion. (Source: http://web.mit.edu/persci/people/adelson/checker shadow_illusion.html)

fourteen-year-old girl. The child fell ill of an unmistakable hysteria, which did in fact clear up quickly and radically under my care. Two months later she died of sarcoma of the abdomen. [29, pp. 197–198]*

Freud's forgetting was due to his failure to diagnose abdominal cancer. Freud was wrong. He diagnosed hysteria, not abdominal cancer. And, allowing his self-concept to get in the way, Freud found it difficult to acknowledge and face up to his error. Freud described hysteria as *unmistakable*, and he recounted his success in the treatment of hysteria. But he could not bring himself to admit the misdiagnosis of cancer.

Freud offers a significant example: showing how reluctance to acknowledge and overcome error encumbers our ability to avoid or to lessen error, and our ability to learn from our mistakes. Freud also demonstrates the example of the accomplished expert who is insufficiently aware of human fallibility. It is a critical fact that error does not discriminate among age, or rank or gender.

7.7 Error and Collaborative Leadership

This book has advanced a collaborative model of leadership, as an idea of *power with*, not an idea of *power over*. In the view of this book, leaders do best when they foster and enrich constructive human exchange.

The collaborative idea recognises that people share responsibility to increase performance, efficiency and safety. Through the collaborative lens, leaders are not superior, special and immune from error. Through the collaborative lens, leaders are responsible and professional. Professionalism does not mean that people cover up mistakes; it means that people try to avoid error, but error is acknowledged and mitigated and used as a basis for learning and improvement, not retribution [25].

The collaborative perspective advanced in this book recognises error as a naturally occurring human phenomenon, which will be overcome *only* by the preparedness to seek understanding and to learn from mistakes. This is important because in organisations defined by a *power over* mindset, leadership is inseparable from notions of hierarchy, and leaders are valued for their swagger and superiority. In organisations like this, making a mistake brings the risk of 'losing face' because leaders are supposed to be 'better than others' and because being 'better than others' is a critical component of institutional success.

In the view of this book, owning up to error should not entail loss of face. Nor should making a mistake be seen as a sign of incompetence. Leaders should be seen as human and professional – not as superior. Responsible, professional people make mistakes just as often as anyone: but rather than

---

* Abridged.

covering up, they 'fess up' and seek to learn from error. And they seek to involve others in their learning.

> There are few single instances that establish a true leader better than one who freely admits a mistake. It demonstrates that all members of the team are truly involved and have a common goal. (This idea is illustrated powerfully by) a senior operations officer whose managerial duties allowed him to keep only minimally qualified in line flying operations. He had a standard introduction to his co-pilots. He always told his new (to him) co-pilot that on this trip the co-pilot's main job was to keep him (the senior officer) out of trouble... (because, despite his seniority)...he was one of the weaker pilots in the system, not getting out to fly as much as he would like. [63, p. 280]*

7.8 A Just Organisation: Non-Jeopardy Reporting

As we have seen, collaborative leaders who understand the human inevitability of error will seek to set a collaborative example. They will also seek to establish supportive institutional systems. Overall, the supportive organisation is known as a *just organisation*. The systems used to enable this *just culture* are known as non-jeopardy reporting systems.

> The aim of non-jeopardy reporting is not to excuse people from bearing the weight of responsibilities; rather it is to encourage people to voluntarily report information that may be critical to identifying the potential precursors to accidents. Following reports, safety issues will be resolved through corrective action. The non-jeopardy reporting system provides for the collection, analysis and retention of safety critical data. When there is an excessive fear of discipline or punishment, this data remains undisclosed. Consequently, risk remains unaddressed, and people remain oblivious. But risk is not disappeared. In fact, in the absence of non-jeopardy reporting systems, risk keeps on escalating. [4]

7.9 Conclusion

Recognising leadership as a collaborative endeavor, this chapter suggests an approach to leadership which focuses on interpersonal communication, situation awareness, decision making and error management. The collaborative approach plays down ideas of leaders as superior. Rather, the collaborative

---

* Slightly abridged.

model demands teamwork and individual responsibility. No one can be excused from wrongdoing because he or she was following orders, or because subordinates did not do as superiors expected. Everyone needs to *realise* what he or she is doing and accept responsibility for his or her actions.

Everyone, in a collaborative environment, must appreciate error and play his or her most responsible part to eliminate error, to mitigate error and to learn from error which slips through the cracks. Leaders, more than anyone, should enable and nurture this sort of constructive human engagement.

8

Leadership and Obligation

8.1 Introduction

In this book, leadership is understood to be a matter of collaboration and relationship, and leaders are understood to be people who exercise constructive influence. Expanding on this theme, the preceding chapters have been informed by the theme of *power with* leadership. The general thrust of argument has tended like this: leaders should be less inclined to use power over others and more inclined to build collaborative cultures.

This argument can give the impression that leaders do not insist on certain things. The impression might be that in being collaborative; leaders are 'soft', ineffectual and hesitant. This is a false impression. Advocating for a collaborative leadership culture, this book recognises that, on occasion, leaders will have to act firmly. On occasion, leaders will require observance of a rule or a standard. Occasionally, leaders must be hard-nosed and uncompromising. On occasion, leaders must put their foot down; they must act rightly, and for right.

But, in the perspective of this book, leadership is not tyrannical; it is not overbearing. Leading is not about overpowering and dominating.

This book recognises leading is not weak and insipid. Leading is not an innocuous sort of friendliness. Leading is hard work. Forging a culture of collaborative joint effort, leaders must mark out boundaries, and they must insist on certain things. To do otherwise would be lax and irresponsible. When they fail to correct noncompliance with critical standards, leaders stand in danger of establishing themselves as the root cause of catastrophe.

Recalling episodes of leadership failures, this chapter makes the point that leaders must act responsibly. There are moments when leaders must make decisions, when they must insist and when indecision and dithering is feckless and wrong.

8.2 Leadership: Power and Failure

The failure of leaders to act properly is often attributable to their overreliance on rules and red tape routine. But rules and procedures are insufficient as a means of avoiding risk or accident. Institutions depend on leaders, especially when the organisation operates in high-risk circumstances. Professor Andrew Hopkins says: 'It is the leaders of an organisation who determine how it functions, and it is their decision-making which determines, in particular, whether an organisation exhibits the practices which go to make up a culture of safety' [41, p. 8].

This book has spoken to the obligations of leaders to foster a collaborative and responsive organisational culture. In the perspective of this book, leaders fail when they use power excessively and when they tolerate or ignore institutional structures and habits which initiate or provoke the exorbitant use of authority.

But leaders fall short as well when they fail to use power responsibly. Leaders fail when they are weak, when they stand for nothing, when they represent nothing, when they become slaves to rules and blind to circumstances.

Rules should regulate only aspects of the organisation that need to be regulated. But when leadership is absent or weak, rules and regulations multiply like bacteria. Customarily pointless, often duplicated and equally often in conflict with each other, uncalled-for rules demonstrate an overreliance on official power and an insufficient awareness of character.

Excessive rules generate colossal administrative friction. They are rules which ornament organisational silos and trap people in oversights. They stifle innovation, and they promote a constricted focus on quantitative measures of compliance and achievement. Excessive and overstated, these rules and the undue reliance on them are symptomatic of a culture of compliance and faintheartedness. Hidden behind rules and psyched up by regulation, people don't have to lead. Overreliant on rules, people lose sight of the obligations of character. They come to rely on procedures, *pro forma* catchphrases and escape clauses: and they lose the habit of standing for anything.

BOX 8.1   THE BUREAUCRACY OF HURRICANE KATRINA

In response to Hurricane Katrina, Federal Emergency Management Agency (FEMA) staff member Douglas Doan struck a deal with operating cafes in New Orleans to supply healthy meals to survivors for $13 a day instead of the $14 being paid for trucked in sandwiches. This was a net saving of US$26,000 per day and supported struggling businesses. The bureaucracy however, 'stopped this common sense arrangement, since it was contrary to a pre-existing policy'. [15]

The failure of people who rely excessively on rules is evident in the reports which follow the bungles. On the public record, official reports allow the bureaucrats to speak for themselves. They reveal the fiasco of the modern establishment, where people fail to stand for anything, fail to insist on anything, fail to own up to anything or bear the responsibility for anything – where, in short, people fail to lead.

8.3 The Official Reports

This section reveals how bureaucratic cunning is a poor substitute for character. Discussion references the report of the Australian Auditor General into the Australian Super Seasprite project [2]. This report exemplifies countless others. It is written in a painstakingly wary style which fails to disguise the bankruptcy of senior people who failed the test of leadership.

Super Seasprite helicopters were acquired for the Navy for the purpose of enhancing the capability of the Navy's eight ANZAC class ships. But no Australian Seasprite helicopter capability exists or ever existed. The Seasprite project was approved in February 1996, with a budget of $746 million Australian dollars. The Navy operated provisionally accepted aircraft between late 2003 and early 2006, when flying was suspended. The project was cancelled in 2008. All in all, expenditure exceeded $1.4 billion Australian dollars.

Yet, despite evident waste and obvious failure, the Australian National Audit Office Report manages to avoid mention of the word leadership. There is no sense of obligation apart from mention of this word in the legal sense of a contract. The word 'wrong' occurs three times in the report. On pp. 260 and 319, it appears in the phrase, 'wrong side of the aircraft', and on p. 327 we read of a 'wrong impression'. Curiously, despite a zealous squandering, despite the non-event that was the Seasprite helicopter, no person was wrong. No person made a mistake. No one was to blame, and neither was anyone responsible or accountable. The word 'blame' appears once in the report, on p. 333, where we read that blame may be set against the 'factors that contributed to the on-going poor performance of the project'.

On the basis of this report we might presume accountability attaches to all things, except people. Indeed, the word 'responsible' appears several times in the report, as a descriptive word in reference to legal or bureaucratic responsibility. But the word is never used in a normative or moral sense.

This report is significant because it is typical of the materially unrevealing and inscrutable reports, which are accepted routinely by the bureaucracy as an explanation. But reports like this are not enlightening; they are not a proper account of reasons projects fail. In the Seasprite report, gnomic phrasing such as 'the failure of the project to provide the required capability', [2]* skirts around the fact that the project was an unequivocal catastrophe. The Seasprite project is described as cancelled, but not failed [2].†

Under the cover of reports like this, bureaucrats in high positions escape scot-free. Their failure to lead is announced by cagey language, which bears out the old adage that the private who loses a rifle suffers greater consequences than the general who loses a war. The official worked out language says nothing plainly. But it speaks clearly to the dearth of moral courage, which undermines institutions. In the failure to speak plainly, and in the disinclination to call perpetrators to account, the official reports locate a crisis of leadership.

In a nutshell, there is a substantial inadequacy to the official reports. Convoluted lawyerly prose sounds severe. But the treatment of serious matters is essentially insincere. Blaming no one, finding the buck stops nowhere, the official reports lift the lid on familiar bad habits of cover up and consequences management. The failure of leadership is plain in the ambivalent language of bureaucratic cunning. The failure is the failure to be plainspoken and to stand up for things that matter.

This failure in moral courage sets the scene for a culture of blind eyes, for a culture where things which should be realised are unseen.

8.4 Blind Eyes

In her analysis of the *Challenger* disaster, Diane Vaughan [85] describes the 'normalisation of evidence', also described as the *normalisation of deviance*. Vaughan offers a sociological explanation of the root causes of accidents. She reveals how leaders come to be spellbound by institutional context: the abnormal comes to be seen as usual.

Launched on 28 January 1986, *Challenger* exploded after 73 seconds. Fragments fell nine miles to the Atlantic. Two solid rocket boosters careened out of the fireball and were destroyed by a range safety officer 110 seconds after launch. All seven crewmembers perished. Stop action film shows the tiny crew compartment intact, freed from its protective encapsulation, tumbling through the smoke cloud. After 2.5 minutes of flight, the crew compartment fell into the sea at 200 miles per hour.

Outwardly, the Rogers Commission concluded the accident was caused by a combustion gas leak through a joint in one of the booster rockets, which was sealed by a device called an O-ring [53, p. 1053]. Cold-soaked while waiting for launch on the pad, the O-ring seal had lost critical flexibility and

---

* Paragraph 9, p. 15. The phrase is repeated at paragraph 1.26 on p. 66.
† Paragraph 10.60, p. 278. The idea of 'project failure' occurs once, in a subheading where the discussion concerns the cost of cancellation.

resilience. Under the pressure and heat of rocket exhaust the O-ring seal dissolved.

But the real cause of the accident was not the unforeseen subito failure of an O-ring seal. The real cause was human, not mechanical. And the real cause had been brewing for years. The real cause was the failure of people who had come to lose proper perspective. Losing proper perspective means that people had come to be unconscious of objectives. In this case, they neglected the urgent need to safeguard life. A blind eye was turned to the astronauts – subliminal, inadvertent and unwitting, perhaps. But that's not the point. These were people trusted to safeguard life, and they were insufficiently attentive to this obligation.

On the night of 27 January 1986, the night before the *Challenger* was due to launch, a three-hour teleconference between Morton Thiokol (manufacturer of the solid rocket motor), the Marshall Space Flight Centre and the Kennedy Space Centre addressed the issue of cold temperature effects on the O-ring seals [17, p. 945]. The participants in this conference observed that at low temperatures, O-ring seals lost critical resilience. But since the O-rings had never failed entirely, the decision was taken to press ahead with the launch schedule.

People were not blind to risk, since they spoke about risk for three hours. But they were careless. They were heedless of the risk they faced. And they were deaf to caution. People chose to press ahead because over time, under-performance of O-ring seals had come to be understood as normal. The parameters of O-ring performance had come to be reconceptualised. And so had levels of risk. Over time, a series of successful cold launches had come to suggest the risk of total O-ring failure was acceptably low.

The disaster 'could have been avoided if only the voice of reasonable caution had been heeded' [31, p. 130]. But this presupposes that the voices of reasonable caution were actually able to speak explicitly and with sufficient gravity and conviction. And this phrase presupposes that the voices of reasonable caution would be heard.

From the *Challenger* accident we learn that the complexity of failure entails that it is impossible to gather all the information, and that there is never one true story of exactly what happened [18, p. 944]. We learn that ideas of safety are not objective and definitive, but rather qualitative and socially constructed [5, p. 268]. Also, and significantly, we see the failure of people who should have been responsible. These were people who allowed self-imposed schedule pressures to blind them to obligations.

The theoretical physicist and Nobel laureate Richard Feynman demonstrated the failure of responsible people. As a member of the Rogers Commission investigating the *Challenger* disaster, Feynman dropped a piece of rubber representing the O-ring seals from *Challenger*'s solid rocket boosters into a glass of ice water. Cold caused the rubber to lose plasticity. Since the O-rings had to stay flexible to work, Feynman's demonstration cut to the heart of the matter [31, p. 131]. Responsible people had turned blind eyes to the dazzlingly obvious.

Feynman demonstrated that people at the elite level of the organisation allowed themselves to behave irresponsibly. People ignored scientific fact and allowed a prevailing culture of mission accomplishment at all costs to dampen their sense of obligation.

This same scenario was played out in the crash of the B-52H Stratofortress, call sign *Czar 52*, on 24 June 1994. As in the loss of *Challenger*, the loss of *Czar 52* was an organisational accident, the consequence of organisational culture – rather than the upshot of a single individual event in a linear causal chain [5, p. 268]. If this culture were to be described, it might be summarised in the words indifference and complacency. The accident reflected the failure of leaders to lead.

The facts, set out by Tony Kern in his book *Darker Shades of Blue* [46], were these. *Czar 52* was launched from Fairchild Air Force Base at approximately 1358 hours to practice for an upcoming airshow. Preparing to land at the end of the practice profile, the crew was required to execute a 'go-around' to avoid another aircraft on the runway. To execute this manoeuvre, *Czar 52* began a 360-degree left turn around the control tower at an altitude of 250 feet above ground level. Three quarters of the way through the turn, the aircraft banked past 90 degrees, stalled, clipped a power line with the left wing and crashed. Impact occurred at approximately 1416 hours. There were no survivors.

The *B-52G/B-52H Pilots' Manual* (T.O. I B-52G-l-l I) published 15 June 1987, recommends, on p. 2-132/AA-2.2, that bank angles should be limited to 30 degrees. The immediate cause of the crash was the action of the pilot in command, Lt. Col. Bud Holland.

But, at a deeper and more meaningful level, this crash was not caused by the error of a pilot, but by the prevailing normalisation of deviant airmanship and lax leadership. In critical ways, the cultivated blind eyes and the misleadership of people who were trusted to act responsibly caused this crash.

8.5 Misleadership

The crash of *Czar 52* revealed the accumulation of deviant tendencies. Rules were bent. Regulations were frequently ignored. Rules proclaimed to be cast iron were actually flexible. Senior people were treated differently from junior people. Standards and expectations were inconsistent. For example, standard training missions were treated differently from evaluations. Likewise, higher headquarters directed missions were seen to be different from ordinary inspections, or airshow demonstrations.

The creep of cultural degradation, and the failure of leadership, was demonstrated by several instances over an extended period of time.

At an airshow at Fairchild Air Force Base in May 1991, Bud Holland was pilot in command of the B-52 display, during which he violated Boeing technical orders by exceeding pitch and bank limits. In addition, he contravened Federal Aviation Regulations, flying directly over the airshow crowd and at lower than permitted minimum altitudes. Senior leaders who watched the display did nothing. Significantly, the people who watched were experienced pilots. Even for them, altitude violations may have been difficult to establish. But pitch and bank angles flagrantly in excess of prescribed limits would have been conspicuous. By their idle leadership, these people encouraged the decay of the squadron culture and sowed the seeds of tragedy.

Two months later, in July 1991, Lt. Col. Holland was pilot in command for the flyover marking a change-of-command ceremony. The flyover plan was developed, briefed and executed without intervention. During practice, and during the flyover itself, Holland accomplished passes below 100 feet. One officer stated that during the ceremony the aeroplane flew so low it blew his cap off. In addition, Holland flew steep bank turns in excess of 45 degrees and at pitch angles in excess of Boeing recommended limits (15 degrees). On one occasion he performed a wingover, a manoeuvre in which the aeroplane is rolled 90 degrees onto its side and the nose allowed to fall through the horizon in order to regain airspeed. The manufacturer prohibits these manoeuvres, because sideslip has potential to cause significant damage.

Following this display, Holland received a verbal warning, but no overt punishment. This is important. It is significant since the issue of a verbal warning indicates that people could see the wrong he was doing. But they were too cowardly to do anything meaningful in response.

On 17 May 1992, Lt. Col. Holland flew the B-52 display at the Fairchild airshow. The display profile included low-altitude steep turns in excess of 45 degrees bank angle and a high-speed pass down the runway. At the completion of this pass, the aeroplane climbed at 60 degrees nose high and levelled off with a wingover. This was the third time in less than a year that Bud Holland had displayed the B-52 with conspicuous disregard for technical airframe limits and general flight safety regulations. Again, nothing was done.

One year later, April 1993, Holland was selected to command a two aircraft 'Global Power' mission to the bombing range in the Medina de Farallons – a small island chain off the coast of Guam. On this mission, he flew in close visual formation to the second aircraft, in contravention of explicit regulations. In addition, also in violation of regulations, and in contravention of common sense, he permitted a crewmember to take video from the bomb bay of live munitions being released from the aircraft. Again, senior leaders did nothing.

Later that same year, in August 1993, Bud Holland was selected as the display pilot for the Fairchild airshow. The display profile was a typical mix: steep turns of greater than 45 degrees of bank, low-altitude passes and a high-pitch manoeuvre estimated to be 80 degrees nose high. Each of these three manoeuvres exceeded technical orders.

On 10 March 1994 – three months before the fatal last flight – Lt. Col. Holland flew a single-aircraft mission to Yakima Bombing Range. On this occasion, the aircraft flew well below the limit of 500 feet. In fact, one pass was photographed at 30 feet above ground level, and on another occasion, the B-52 was estimated to be in the neighbourhood of three feet above the ground. Spiritless leadership enabled Holland's flagrant and reckless impetuosity.

People were not blind to Holland's deviance. People could see the wrong he was doing, since they issued him with reprimands. But the decision to ground Lt. Col. Holland was too big. A reprimand was less than hell to pay; it was a mere formality. It was just enough for people to say that Holland had been punished. But it was a long way from convincing, and a long way from helpful. Responsible people made official representations; they sought real action to stop a dangerous pilot. But senior people, in possession of official power, were too small to suspend Holland from flying. They were too timid to use their power responsibly.

Faced with a pilot whose unquestioned stick and rudder skill eliminated the chance that his overbanks and excess pitch angles were unintended errors, senior leaders did nothing. Bud Holland contravened explicit safety criteria, and senior leaders did nothing – they were enthralled by his charisma and in awe of his physical skill.

Bud Holland had accumulated more hours than anyone in the B-52. But he had also become dangerous. He was complacent, reckless and a wanton violator of rules. Tragically, the people who should have grounded Holland tolerated him, like the eccentric aunt about whom people say, 'Just ignore her'.

8.6 Conclusions

Blessed with non-involvement and hindsight, it is easy for retrospective observers to wonder at how people could be foolish, arrogant or reckless. But looking ahead, we must learn from the mistakes of the past. We must own the blunt truth: people do no good when they allow themselves to behave irresponsibly.

Leaders fail when they fail to establish a healthy organisational culture. Confronted with evidence of regulatory deviations, leaders must find the courage of their convictions; they must face the challenges they are obligated to face. Leaders must insist on certain things. They must use power wisely and act when to remain idle would be negligent or delinquent. Leaders must rise above the weak-willed tangle of bureaucracy. They must stand apart from the unclear mainstream organisational maze; they must provide direction and lucidity.

In the perspective of this book, *the way the leader acts* is critical. Leaders do not use power wantonly, but astutely. When they act with any degree of firmness, their actions are characterised by justice and fairness. In the perspective of this book, a leader is first among equals, committed to the ideal of *power with*, and obligated to act rightly and for right.

9

Conclusion

This book has argued against what we might call an old-style command-and-control approach to leadership. In this book, rather than the power of one *over* others, leadership has been seen to be most constructive when it is understood as the collaborative power of one *with* others.

The idea of 'power with' was described as especially helpful in safety-critical and high-risk domains. In these fields, directive styles of command and control are orthodox. But high-authority, crack-the-whip approaches are not as constructive as they are often assumed to be. In this book we explained that in a safety-critical domain, such as medicine, aviation, or the fire and emergency services, people face challenging situations. Decisions need to be made with ambiguous and often conflicting information in circumstances of intense strain. Pressure comes from fatigue and stress, from inadequate resources and from time limits. Working environments might be dangerous, made complex by technology and difficult by regulation. The consequences of mistake are extreme and often beyond measure in the incalculable terms of human life.

These are circumstances in which leaders need to lead. In the perspective of this book, when regulation, direction or firmness is required, leaders should exert only so much authority as to ensure tasks are properly carried out. The real power in leadership is the power of the collective; the power leaders build with others in a joint effort. Leaders do best when they build collaborations and foster the sort of mutual institutional culture that lets people do well.

9.1 The Contribution of This Book

Typically, when we look to leadership in emergencies, disasters or incidents, we look to leadership *in the moment*. We look to see someone 'in charge', taking responsibility for coordination and decision.

But the truly significant influence of leadership precedes the critical incident by years. Leadership establishes the foundational elements of organisational or institutional culture.

Culture defines background conditions which are critical, as immersed in the institutional setting people learn to do well or poorly. People exist in contexts. People cannot be wholly or properly understood in isolation.

Asserting the power and influence of the organisational context, this book has suggested that to understand people, and the way people respond in moments of crisis, we must understand the way people are led in moments of calm.

In an emergency, when the stakes are high, when people need to be counted on, it is important that people are accustomed to thinking for themselves, to acting with initiative, to getting along with others, to admitting mistakes and correcting them. In an emergency it is not constructive when people are daunted by the boss and are timid, anxious and afraid to act. It is not constructive when people are acculturated to stifling bureaucratic compliance. It is not helpful when people wait to be told what to do, when they have a check box mentality, when they are afraid of consequences, when they cannot make a decision to save themselves.

9.2 Beyond Command and Control

Advancing a collaborative idea of leadership, this book has challenged the misbelief that leadership is commanding, controlling or dominating. It is a mistake to think senior people know it all, that they never make a mistake, and that the rest of us have to wait for the big fish to tell us what to do. But this misreading of leadership has a very wide currency and resonance. This is a misreading which favours the attention-seeker over the expert and the bully over the consensus-builder. Unpleasant at the best of times, this is a misinterpretation which may prove fatal in the safety-critical domain.

Leadership's best effect is in the engagement of people.

9.3 Collaborative Culture

Looking beyond command and control, this book has asked leaders to be mindful of their larger responsibility to build a communicative, collaborative culture. Even with clear goals and adequate resources, experienced and capable people can fail in the heat of the moment because they can be brought undone by the sort of breakdown in coordination that arises from miscommunication or interpersonal conflict.

Leadership which leads to failure in the critical moment is pervasive. It is cashed out in the habits and the assumptions which define institutions and the risk and safety systems, which extend beyond single institutions and agencies. It is important because we face emergencies with the institutions we have, not the organisations or cultures we would like – or which we need.

Thus, leaders need to make it possible for people – including people from different organisations and from different jurisdictions – to work together.

Successful leaders will foster the sort of culture within which people at the coalface feel free to exercise initiative and common sense. Thus, in this book, constructive leadership has been described as the power leaders evolve *with* others.

9.4 Power With

Coined by Mary Follett in her important book, *Creative Experience*, the idea of 'power with' was a cornerstone of this text. Foreshadowing the concept of soft power advanced by Joseph Nye of Harvard University, the idea of 'power with' speaks against the polarity implied in terms such as 'powerful and powerless' or 'superior and subordinate'.

The 'power with' approach enables the organic collaborative and innovative responsiveness of the group. When there is 'power with' there is effective partnership, actual communication and reduced risk. In a climate of 'power with', people are capacitated to speak against error. And they are empowered to decide, to use their initiative and act. Obviously, mistakes may be made. But the small risk of error is better than the high probability of indecision. And, since people are typically well trained, errors will be occasional rather than likely.

Another way of describing the 'power with' approach to leadership was offered in the significant idea of the leadership 'system'. The idea of a system was explained by a nonsense question: What would you rather have, the head or the heart? The question is silly, since we cannot have one or the other. We must have each together. The body is a system – it works only when it is interconnected. Each part has a specific function, and each part depends on the other elements.

So it is in the case of 'power with' leadership. The organisation, the team or the interagency task group is a system: leading and following is a partnership. The point is: positional seniority is important, but positional seniority (misdescribed as leadership) is not more important than the connection and interaction between people, which is the source and engine of institutional efficiency. The directive, emphatic and assertive habits of 'command and control' have a place. But they almost inevitably have unfortunate effects when they outrun ideas of integration.

9.5 The Leaders Who Do Best

Power is very easily overplayed, and when overstated it has a destructive effect: building barriers and resentments and wrecking the ability of leaders to nourish partnerships and combination.

This book was not an argument against power. It argued against the abuse or the overstatement of power. The book asked for a new orientation toward power, for a subtler and more nuanced reading of power.

Doing so, this book recognised that the leaders who do best build an organisational culture which recognises the value of teamwork, and they train people to develop shared situation awareness about evolving events. The leaders who do best recognise that when complex situations are unfolding rapidly, what matters is adaptive coordination among team members. This sort of performance arises from collaborative leadership – not bullish command and control.

In a nutshell, leaders set people up for success when they foster the collaborative culture of partnership, when they foster the readiness to be flexible, to be responsive and to be cooperative. Leaders prepare their people by leading, not by dominating. In the perspective of this book, when leaders have acculturated people to bureaucratic subservience and unthinking compliance they have deprived them of tactical independence and, as leaders, they have failed.

9.6 Command-and-Control Structures

Bureaucratic hierarchies are necessary because they offer critical exactness to legal authority and accountability. This sort of explicit accountability is an essential safeguard against malfeasance and shady dealing. But when bureaucracy becomes undue and overprominent, and when there is an overemphasis on institutional hierarchy, then people lose their energy and innovative spark.

Besides legal accountability, hierarchical command-and-control structures enable efficiency. Bureaucratic structures enable regulation, exactness, precision, and repeatability. Thanks to bureaucratic control, tasks can be done with increasing efficiency and reduced waste, over and over again. But hierarchy and bureaucracy are insufficient for leadership.

In this book we cautioned readers not to confuse command and control for leadership. And we asked people to consider how the behaviours associated with constructive influence can be inhibited and counteracted by an overprominent bureaucracy.

The point is: leaders working in organisational structures need to be mindful of the undermining effects of command-and-control bureaucracy. This is important because, if not checked, bureaucracies establish the conditions for poor performance and elevated risk.

9.7 Compelling, Not Coercive

This book has depicted a view of leadership which is all about the value of individuals, and the advantage that arises from collaboration. In the perspective of this book, leadership does not derive from positional power, from formal authority or from standing in an institutional hierarchy. Leaders encourage people, and they enable people to join forces and to participate as responsible individuals in an institutional enterprise.

The other way, the bureaucratic way, expects people to play their little part, doing as they are told, abiding methodically by the job description. But in the view of this book, leaders should enable people to derive human fulfilment from work. This is satisfaction which comes from intelligent co-participation. In the view of this book, people are more than things who serve the ends of the boss, and leadership is more than bureaucratic boss-ism. In the perspective of this book, leaders bring strength of character, professional values and human ideals to formal positions.

In this way, leaders bring personality, human understanding and judgement to formal positions. As a richly collaborative human endeavour, leadership is compelling, not coercive.

Leadership defined by character and right purpose *co-opts*, attracting others to follow willingly. This remark presumes that one person cannot lead another *unless the other accepts that relationship*. For this reason, leadership does not require rank, but it does demand character, empathy and the readiness to come together in constructive partnership with others.

In writing this book we believe that traditional overbearing, bossy cultures need to be challenged. We believe that people will develop the habits they will need in times of trial only when they are well led. We hope that the discussion and resources in this book will help leaders build collaborative partnerships that move beyond the established orthodoxies of command and control.

References

1. National Transportation Safety Board. Aircraft accident report: NTSB-AAR-79-7. Washington, DC: NTSB, 1979.

2. Attorney-General's Department, The Commonwealth of Australia Department of Defence and the Australian National Audit Office. The Super Seasprite: The auditor-general audit report no. 41 2008-09 performance audit. 2009.

3. Driskell, J. and R. Adams. *Crew resource management: An introductory handbook*. Washington, DC: US Department of Transportation, Federal Aviation Administration, 1992.

4. Flight Standards Service, Federal Aviation Administration, (William J. White). Advisory circular 120-66: Aviation Safety Action Programs (ASAP), 1997; **1**: 8.

5. Atak, A. and S. Kingma. Safety culture in an aircraft maintenance organisation: A view from the inside. *Safety Science*, 2011; **49**(2): 268–78.

6. Barton, M. A. and K. M. Sutcliffe. Overcoming dysfunctional momentum: Organizational safety as a social achievement. *Human Relations*, 2009; **62**(9): 1327–56.

7. Besnard, D. and E. Hollnagel. I want to believe: Some myths about the management of industrial safety. *Cognition, Technology & Work*, 2014; **16**(1): 13.

8. Brim, O. G. *Personality and decision processes*. Studies in the Social Psychology of Thinking, Vol. 2. Stanford, CA: Stanford University Press, 1962.

9. Bruns, A., J. Burgess, K. Crawford and F. Shaw. #qldfloods and @qpsmedia: Crisis communication on twitter in the 2011 south east Queensland floods. Brisbane: ARC Centre of Excellence for Creative Industries and Innovation, 2012.

10. Cannon-Bowers, J. and E. Salas. Decision making under stress: Theoretical underpinnings. In J. Cannon-Bowers and E. Salas (Eds.), *Decision making under stress: Implications for individual and team training* (pp. 17–38). Washington, DC: American Psychological Association, 1998.

11. Cannon-Bowers, J. A. and E. Salas. (Eds.) *Decision making under stress: Implications for individual and team training*. Washington, DC: American Psychological Association. 1998.

12. Carroll, J. S. and B. Fahlbruch. The gift of failure: New approaches to analyzing and learning from events and near-misses. Honoring the contributions of Bernhard Wilpert. *Safety Science*, 2011; **49**(1): 1–4.

13. Chabris, C. F. and D. J. Simons. *The invisible gorilla: Thinking clearly in a world of illusions*. New York: HarperCollins, 2010.

14. Cook, P. J. and B. Smith. Leading innovation, creativity and enterprise. *Industrial and Commercial Training*, 2016; **48**(6): 294–9.

15. Cooper, C. and R. Block. *Disaster: Hurricane Katrina and the failure of Homeland Security*. New York: Macmillan, 2007.

16. Curnin, S., C. Owen, B. Brooks, D. Paton and C. Trist. Role clarity, swift trust and multi-agency coordination. *Journal of Contingencies and Crisis Management*, 2015; **23**(1): 29–35.

17. Dalal, S. R., E. B. Fowlkes and B. Hoadley. Risk analysis of the space shuttle: Pre-Challenger prediction of failure. *Journal of the American Statistical Association*, 1989; **84**(408): 945–57.

18. Dekker, S., P. Cilliers and J.-H. Hofmeyr. The complexity of failure: Implications of complexity theory for safety investigations. *Safety Science*, 2011; **49**(6): 939–45.

19. Dracup, K. and C. W. Bryan-Brown. From novice to expert to mentor: Shaping the future. *American Journal of Critical Care*, 2004; **13**(6): 448–50.

20. Duhigg, C. *The power of habit: Why we do what we do in life and business.* New York: Random House, 2012.

21. Edmondson, A. Psychological safety and learning behavior in work teams. *Administrative Science Quarterly*, 1999; **44**(2): 350–83.

22. Edmondson, A. C. Speaking up in the operating room: How team leaders promote learning in interdisciplinary action teams. *Journal of Management Studies*, 2003; **40**(6): 1419–52.

23. Edmondson, A. C. Learning from failure in health care: Frequent opportunities, pervasive barriers. *Quality and Safety in Health Care*, 2004; **13**(suppl 2), ii3–9.

24. Endsley, M. R. Situation Awareness in Dynamic Human Decision Making: Theory and Measurement. A dissertation presented at the Faculty of the Graduate School, University of Southern California, Los Angeles, California, pp. 9–10, May 1990.

25. Ernsting, J., A. N. Nicholson and D. Rainford. *Aviation Medicine*. London: Arnold, 2003.

26. Fielder, J. H. and D. Birsch (Eds.). *The DC-10 case: A study in applied ethics, technology, and society.* Case Studies in Applied Ethics, Technology, and Society. Albany: State University of New York Press, 1992.

27. Flin, R. H., P. O'Connor, and M. Crichton. *Safety at the sharp end: A guide to non-technical skills.* Farnham, UK: Ashgate, 2008.

28. Follett, M. P. *Creative experience.* New York: Longmans, Green & Co., 1924.

29. Freud, S. *Group psychology and the analysis of the ego.* New York: W. W. Norton, 1975.

30. Galotti, K. M. *Making decisions that matter.* Mahwah, NJ: Lawrence Erlbaum, 2002.

31. Grossman, K. Book review. *Truth, lies, and o'rings: Inside the Space Shuttle Challenger disaster*, by Allan J. Mcdonald with James R. Hansen. *Science Communication*, 2010; **32**(1): 130–2.

32. Hackett, J. *A leader – Not a paragon.* In *Serve to lead*. Sandhurst, UK: Royal Military Academy, n.d.

33. Hackman, J. R. *Leading teams: Setting the stage for great performances.* Brighton, MA: Harvard Business Publishing, 2002.

34. Hawkins, F. H. and H. W. Orlady. *Human factors in flight.* Aldershot, UK: Ashgate, 2000.

35. Helmreich, R. L. Managing human error in aviation. *Scientific American*, 1997; **276**(5): 62–7.

36. Helmreich, R. L. and A. C. Merritt (Eds.). *Culture at work in aviation and medicine: National, organizational, and professional influences.* Brookfield, VT: Ashgate, 1998.

37. Helmreich, R. L., A. C. Merritt and J. A. Wilhelm. The evolution of crew resource management training in commercial aviation. *The International Journal of Aviation Psychology*, 1999; **9**(1): 19–32.

38. Hofstede, G. H. Culture and organizations. *International Studies of Management & Organization*, 1980; **10**(4): 15–41.

39. Hofstede, G. National cultures in four dimensions: A research-based theory of cultural differences among nations. *International Studies of Management & Organization*, 1983; **13**(1–2): 46–74.

40. Hofstede, G. *Culture's consequences: International differences in work-related values.* Thousand Oaks, CA: SAGE, 1984.

41. Hopkins, A. *Safety, culture and risk: The organisational causes of disasters.* Sydney: CCH Australia Ltd., 2005.

42. Jackson, C. et al. Dynamics of a memory trace: Effects of sleep on consolidation. *Current Biology*, 2008. **18**(6): p. 393–400.

43. Janis, I. *Groupthink*, 2nd ed. Boston: Wadsworth, 1982.

44. Job, M. *Air disaster*, Vol. 1. Canberra: Aerospace Publications, 2004.

45. Kanki, B. G., R. L. Helmreich and J. M. Anca. *Crew resource management.* Academic Press/Elsevier, 2010.

46. Kern, T. T. *Darker shades of blue: The rogue pilot.* New York: McGraw–Hill.

47. Kerstholt, J. H. and J. G. Raaijmakers. Decision making in dynamic task environments. In R. Ranyard, W. R. Crozier, and O. Svenson (Eds.), *Decision making: Cognitive models and explanations* (pp. 205–17). London and New York: Routledge, 1997.

48. Klein, G. A. A recognition-primed decision (RPD) model of rapid decision making. In G. A. Klein, J. Orasanu, R. and R. Caldenwood, *Decision making in action: Models and methods.* Cognition and Literacy. New York: Ablex, 1995.

49. Langewiesche, W. *Columbia's* last flight. *The Atlantic Monthly*, 2003; **292**(4): 73.

50. Lewis, A., T. E. Hall and A. Black. Career stages in wildland firefighting: Implications for voice in risky situations. *International Journal of Wildland Fire*, 2011; **20**(1): 115–24.

51. Lipshitz, R., G. Klein, J. Orasnu and E. Salas. Taking stock of naturalistic decision making. *Journal of Behavioral Decision Making*, 2001; **14**(5): 331–52.

52. Lyng, S. Edgework: A social psychological analysis of voluntary risk taking. In *American Journal of Sociology*, 1990; **95**(4): 851–86.

53. Maranzano, C. J. and R. Krzysztofowicz. Bayesian reanalysis of the *Challenger* O-ring data. *Risk Analysis*, 2008; **28**(4): 1053–67.

54. Marcus, L. J., B. C. Dorn and J. M. Henderson. Meta-leadership and national emergency preparedness: A model to build government connectivity. *Biosecurity and Bioterrorism: Biodefense Strategy, Practice, and Science*, 2006; **4**(2): 128–34.

55. McIntosh, S. E. The wingman-philosopher of MiG Alley: John Boyd and the OODA Loop. *Air Power History*, 2011; **58**(4): 24–33.

56. Meyerson, D., E. W. Karl and R. M. Kramer. Swift trust and temporary groups. In R. M. Kramer and T. R. Tyler (Eds.), *Trust in organizations: Frontiers of theory and research* (p. 167). Thousand Oaks, CA: SAGE.

57. Miyagi, M. *Serious accidents and human factors: Breaking the chain of events leading to an accident: Lessons learned from the aviation industry.* Reston, VA: American Institute of Aeronautics & Astronautics, 2005.

58. Morrison, J. G., R. T. Kelly, R. A. Moore and S. G. Hutchins. Implications of decision-making research for decision support and displays. In J. A. Cannon-Bowers and E. Salas (Eds.), *Making decisions under stress: Implications for individual and team training* (pp. 375–406). Washington, DC: American Psychological Association, 1998.

59. Newell, A. and H. A. Simon. *Human problem solving.* Englewood Cliffs, NJ: Prentice-Hall, 1972.

60. Nye, J. S. Power and leadership. In N. Nohria and R. Khurana (Eds.), *Handbook of leadership theory and practice: A Harvard Business School centennial colloquium.* pp. 305–32. Brighton, MA: Harvard Business Publishing, 2010.

61. Orasanu, J. Stress and naturalistic decision making: Strengthening the weak links. In R. Flin, E. Salas, M. Strub and L. Martin, *Decision making under stress: Emerging themes and applications* (pp. 43–66). Aldershot, UK: Ashgate, 1997.

62. Orlady, H. W. and C. Foushee (Eds). Cockpit resource management training: Proceedings of a workshop conducted by NASA Ames Research Center and the US Air Force Military Airlift Command, San Francisco, California, May 6–8, 1986, NASA Technical Report No. CP-2455. Washington, DC: NASA Scientific and Technical Information Branch, 1987.

63. Orlady, H. W. and L. M. Orlady. *Human factors in multi-crew flight operations.* Aldershot, UK: Ashgate, 1999.

64. Owen, C. Leadership, communication and teamwork in emergency management. *Human factors challenges in emergency management: Enhancing individual and team performance in fire and emergency services* (p. 125). Aldershot, UK: Ashgate, 2014.

65. Patterson, J. M. and S. A. Shappell. Operator error and system deficiencies: Analysis of 508 mining incidents and accidents from Queensland, Australia using HFACS. *Accident Analysis & Prevention*, 2010; **42**(4): 1379–85.

66. Piattelli-Palmarini, M. *Inevitable illusions: How mistakes of reason rule our minds.* New York: John Wiley & Sons, 1994.

67. Pilcher, J. J. and A. J. Huffcutt. Effects of sleep deprivation on performance: A meta-analysis. In *Sleep: Journal of Sleep Research & Sleep Medicine*, May 1996; **19**(4): 318–26.

68. Pinker, S. Words and rules in the human brain. *Nature*, 1997; **387**: 547–8.

69. Reason, J. *Human error.* Cambridge: Cambridge University Press, 1990.

70. Reason, J. and A. Hobbs. *Managing maintenance error.* Aldershot, UK: Ashgate, 2003.

71. Ruffell-Smith, H. A simulator study of the interaction of pilot workload with errors, vigilance, and decisions (technical memorandum 78482). Hampton, VA: NASA Scientific and Technical Information Office, 1979.

72. Sagan, S. D. *The limits of safety.* Princeton, NJ: Princeton University Press, 1993.

73. Schultz, K. *Being wrong: Adventures in the margin of error.* New York: HarperCollins, 2010.

74. Shappell, S. A. and D. A. Wiegmann. Applying reason: The human factors analysis and classification system (HFACS). *Human Factors and Aerospace Safety*, 2001; **1**(1): 59–86.

75. Sherman, P. J., R. L. Helmreich and A. C. Merritt. National culture and flight deck automation: Results of a multination survey. *International Journal of Aviation Psychology*, 1997; **7**(4): 311–29.

76. Simon, H. A. Rational choice and the structure of the environment. *Psychological Review*, 1956; **63**(2): 129.

77. Simpson, D. Thinking about IED warfare: Defeating the insurgent weapon of choice. *Marine Corps Gazette*, 2011; **95**(9): 35–41.

78. Smith-Jentsch, K. A., J. H. Johnston and S. C. Payne. *Measuring team-related expertise in complex environments.* In J. A. Cannon-Bowers & E. Salas (Eds.), *Decision making under stress: Implications for individual and team training* (pp. 61–87). Washington, DC: American Psychological Association.

79. Snook, S. *Friendly fire: The accidental shootdown of U.S. Black Hawk helicopters over Norther Iraq.* Oxford: Princeton University Press.

80. Stanton, N. A., P. M. Salmon, G. H. Walker and D. P. Jenkins. Is situation awareness all in the mind? *Theoretical Issues in Ergonomics Science,* 2010; **11**(1–2): 29–40.

81. Storr, J. A command philosophy for the information age: The continuing relevance of mission command. *Defence Studies,* 2003; **3**(3): 119–29.

82. Tannenbaum, S. I., K. A. Smith-Jentsch and S. J. Behson. Training team leaders to facilitate team learning and performance. In J. A. Cannon-Bowers and E. Salas (Eds.), *Making decisions under stress: Implications for individual and team training* (pp. 247–70). Washington, DC: American Psychological Association, 1998.

83. United States Army Annual Report. The army profession: After more than a decade of conflict. In *United States Army training and doctrine command.* U.S. Army: Authorised for Distribution by Robert W. Cone, General, 2012.

84. Van Dongen, H. P. A., G. Maslin, J. M. Mullington and D. F. Dinges. The cumulative cost of additional wakefulness: Dose-response effects on neurobehavioral functions and sleep physiology from chronic sleep restriction and total sleep deprivation. In *Sleep,* 2003; **26**(2): 117–265.

85. Vaughan, D. *The Challenger launch decision: Risky technology, culture, and deviance at NASA.* Chicago: University of Chicago Press, 1996.

86. Weick, K. E. The collapse of sensemaking in organizations: The Mann Gulch disaster. *Administrative Science Quarterly,* 1993; **38**(4): 628–52.

87. Weick, K. E. *Making sense of the organization,* Vol. 2: *The impermanent organization.* Hoboken, NJ: John Wiley & Sons, 2012.

88. Wiegmann, D. A. and S. A. Shappell. Human error analysis of commercial aviation accidents: Application of the human factors analysis and classification system. *Aviation, Space and Environmental Medicine,* 2001; **72**(11): 1006–16.

89. Witte, E., N. Joost and A. L. Thimm. Field research on complex decision-making processes: The phase theorem. *International Studies of Management & Organization,* 1972; **2**(2): 156–82.

Index